"There is a magic in the interior of this island that few will write about or speak of to others – an endless fascination with the land. Gary Collins is entranced in the same way that the allure of rock, tree, and bog seized the indomitable Allan Keats, and before him, his ancestor, the Mi'kmaq Soulis Joe. This book gives voice not only to these men but to the great and wonderful wilderness of Newfoundland. Read it and be prepared for the wonder and love of the wild places. It will grab and hold on to you, too."

J. A. RICKETTS, AUTHOR OF
THE BADGER RIOT

"To quote from *Soulis Joe's Lost Mine*, I have come to love it all – every page and every paragraph that Gary Collins uses to portray the life and soul of prospecting, and the love of the land that the Mi'kmaq showed so naturally."

AAMIE GILLAM

"Gary Collins simply gets better and better. This is his best."

GARRY CRANFORD, AUTHOR OF
NEWFOUNDLAND SCHOONER:
NORMA & GLADYS

SOULIS JOE'S LOST MINE

A NEWFOUNDLAND MEMOIR

GARY COLLINS

FLANKER PRESS LIMITED
ST. JOHN'S
2009

Library and Archives Canada Cataloguing in Publication

Collins, Gary, 1949-
Soulis Joe's lost mine : a memoir / Gary Collins.

ISBN 978-1-897317-38-9

1. Soulis Joe. 2. Micmac Indians--Newfoundland and Labrador. 3. Prospecting--Newfoundland and Labrador. 4. Newfoundland and Labrador--History. I. Title.

FC2173.1.S57C65 2009 970.004'97343 C2009-903714-9

PRINTED IN CANADA

Cover design: Adam Freake
"Sourdough" illustration by Clint Collins

Some racial epithets appear in this book to reflect
the prejudices of the characters who use them.

FLANKER PRESS
PO BOX 2522, STATION C
ST. JOHN'S, NL, CANADA
TOLL-FREE: 1-866-739-4420
WWW.FLANKERPRESS.COM

14 13 12 11 10 09 1 2 3 4 5 6 7 8 9

Canada

Canada Council for the Arts Conseil des Arts du Canada

Newfoundland Labrador

We acknowledge the financial support of the Government of Canada through the Book Publishing Industry Development Program (BPIDP) for our publishing activities; the Canada Council for the Arts which last year invested $20.1 million in writing and publishing throughout Canada; the Government of Newfoundland and Labrador, Department of Tourism, Culture and Recreation.

If the reader will allow, I will make three dedications in this book. First, I respectfully dedicate these pages to the Mi'kmaq people of Newfoundland and Labrador, whose contribution to the progress of the European exploration of this land is unequalled in the annals of any native peoples, anywhere. Second, to the true Red Indian, the Beothuk, whose spirit I found still moving through the shadowed glades above the shores of the huge lake which still bears their name. Third, and with much respect, to Allan Keats, a consummate woodsman, proud advocate of his Indian heritage, and ever a searcher of mysteries. I am proud to call him my friend.

Also by Gary Collins

Cabot Island
The Last Farewell

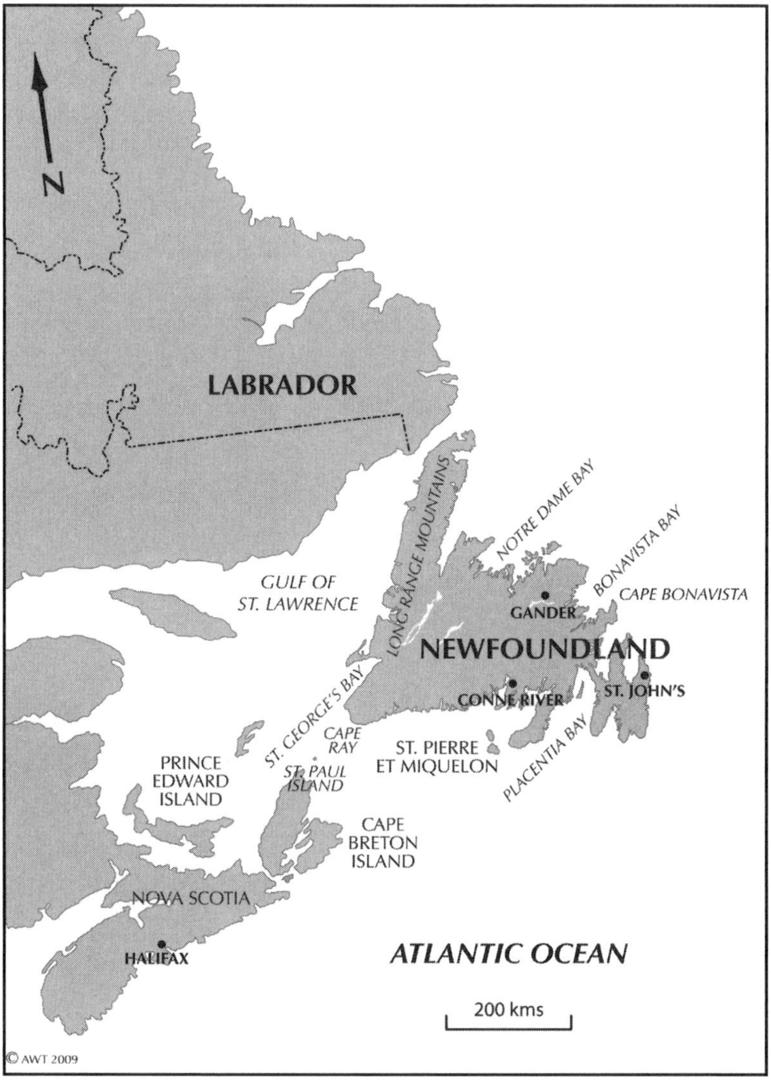
N
LABRADOR
GULF OF
ST. LAWRENCE
LONG RANGE MOUNTAINS
NOTRE DAME BAY
BONAVISTA BAY
CAPE BONAVISTA
GANDER
NEWFOUNDLAND
CONNE RIVER
ST. JOHN'S
ST. GEORGE'S BAY
CAPE
RAY
ST. PIERRE
ET MIQUELON
PLACENTIA BAY
PRINCE
EDWARD
ISLAND
ST. PAUL
ISLAND
CAPE
BRETON
ISLAND
NOVA SCOTIA
HALIFAX
ATLANTIC OCEAN
200 kms
© AWT 2009

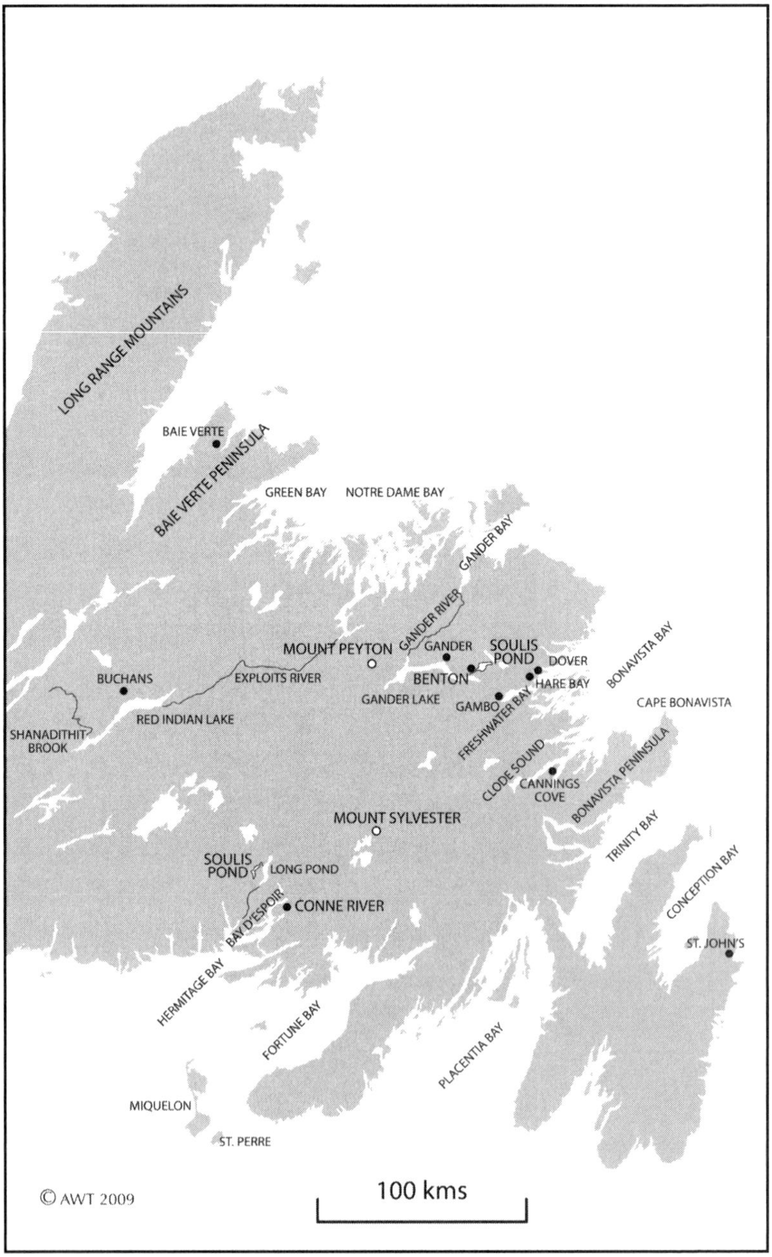

LONG RANGE MOUNTAINS
BAIE VERTE
BAIE VERTE PENINSULA
GREEN BAY
NOTRE DAME BAY
GANDER BAY
GANDER RIVER
MOUNT PEYTON
GANDER
SOULIS POND
DOVER
BENTON
HARE BAY
BONAVISTA BAY
BUCHANS
EXPLOITS RIVER
GANDER LAKE
GAMBO
CAPE BONAVISTA
RED INDIAN LAKE
FRESHWATER BAY
SHANADITHIT BROOK
CLODE SOUND
CANNINGS COVE
BONAVISTA PENINSULA
MOUNT SYLVESTER
TRINITY BAY
SOULIS POND
LONG POND
CONCEPTION BAY
CONNE RIVER
BAY D'ESPOIR
ST. JOHN'S
HERMITAGE BAY
FORTUNE BAY
PLACENTIA BAY
MIQUELON
ST. PERRE
© AWT 2009
100 kms

Prologue

I met him at a cancer relay function. He and his wife, Christina, their two sons and daughter were walking for Kevin's mother, Norma. I was walking for my wife, Rose. They were both cancer survivors. The rain was pouring, making the raw June 13 night even darker than normal.

"Kevin Keats," he said, reaching out a big hand. We shook firmly.

Don't look much like an Indian, I thought. *Big bugger, too*.

We talked for a while about the terrible cancers that seemed to seek out and invade every family. He wore a trendy below-the-waist-length jacket and a ridiculous-looking woollen hat on his head, complete with tassels that drooped below his earlobes. His eyes were dark above high cheekbones, deep-set and capable of compassion as well as dogged determination. On second glance, I decided he could be Indian after all.

We were under a huge tent erected by the town of Gambo for their very first Cancer Walk for Life event. "I read your two books," he said as we waited for the first part of the walk to begin. "I like your style of writing. Especially how you portray our own way of life and language. I really enjoyed them both. Good job."

I thanked him for the compliment, a bit uncomfortable as always when praised for my writing ability. I could get used to it, though.

"I'd love to write a book about my family history, too, but hell, I could never find the time. How about you writing it for me?" He was very direct.

"You mean ghostwrite it?" I asked, surprised at the request.

"I'm not sure what that means," he replied.

I explained I would do the actual writing, but the book would carry his name.

"Not a chance in hell!" he retorted right away. "I'll take no credit for something I didn't do. I'd like for you to write the book, bring my great-great-grandfather, the old Indian, Soulis Joe, alive. Tell the story of our family history as it relates to the Mi'kmaq nation, like you did with the people in your other books."

I knew he was involved in the mining industry, particularly in the prospecting part of it. My son-in-law, Mike, was working for him somewhere in Nunavut, looking for uranium.

The Keats family lived in the tiny hinterland town of Benton. They were the sixth generation of Newfoundland prospectors known the world over for their ability to find ore bodies and precious metals. Everyone in our area knew about the Indian prospectors from Benton. They were of Mi'kmaq Indian descent and had just received an international award for their abilities. This much I knew about them. Nothing else. I told him I would be happy to give him any assistance I could in compiling a manuscript of his own, but that I was already under contract with my publisher for another book.

"I could probably do the book," Kevin replied, "but I just don't have the time, b'y. You think about it some more and let me know."

We talked some more about his business. He was very excited about his work. It was kind of contagious to hear him go on about the exploration for minerals all over the world, especially here in our province. It was his passion, I could tell.

Still, I knew nothing about minerals at all. Not the geology part of it, about faults and sliding masses of whole continents and volcanic eruptions, least of all where to look for minerals. Besides, I had to tell him, I didn't think a book about prospecting would be a very good seller. I thought it would be a bit boring, actually, but I never told him that.

"You'd be surprised at how many people would be interested in what's going on," he replied. "The world is on a 'mineral high' right now."

From outside the crowded tent a voice crackled through the rain-drenched loudspeaker and informed us it was time to line up for the first walk. Kevin strode away.

"Give it some thought," he said in parting, and disappeared with the crowd out into the driving rain. We met once more that miserable night but spoke no more about the idea of a book.

Over the next few days, I gave the idea of writing about the Mi'kmaq prospectors some consideration. I figured I could write the book if I was given enough information. But how would I get it? I'm a stickler for detail and won't write anything without having complete knowledge of it. It was then I had a wild idea. Still, even if my idea got Kevin's approval, it would need the complete endorsement of the most important player of all, my publisher, Flanker Press. After all, it was one thing to write a book – I shuddered at the amount of research involved – but quite another to convince a publisher, who would be expected not only to promote and sell the book but also foot the production costs, as well as take care of the actual distribution and sales.

I called my publisher and got my first surprise. Garry Cranford of Flanker Press had spent five years of his earlier life prospecting. He knew first-hand that it was very interesting. Still, a book about prospecting? Probably not very interesting to the general public. He voiced my own feelings.

We talked awhile. I told him about the idea I had. I also informed him this was a family of Mi'kmaq Indian descent, who had more than a hundred years' involvement with prospecting, much of it for major mining companies. I also informed him they were known the world over in mining circles. He already knew about their recent award.

I also ran by him the way in which I would like to do the work. It wouldn't be a document about minerals and metals and mines, but a brought-to-life story about the search for minerals, both the old way and the new.

My second surprise: "Tell you what. You write it in your style of telling stories. Tie in some history of the Mi'kmaq people on the island, and if I can get this guy, Kevin Keats, to take a few copies of the book to his conventions, we'll publish it."

And that was that.

"About the book you want written," I said to Kevin when I finally got him on the phone several days later. He has few idle moments. "Believe it or not, my publisher used to be a prospector himself. He will publish the book if I do it in my style. You know what I mean. Walking, talking, living the life of a hands-on prospector."

Kevin was very pleased. There was nothing he couldn't tell me about prospecting, he said.

"Ah – as to that part of it," I interrupted, "here's my idea. Hire me for the summer, pay me the wage of your other guys in the field, and put me as a partner with someone who knows the prospecting end of your business, as well as your family history, if possible. I'll get a summer's wage, learn about your work first-hand, and pick up your family history. I'll get the royalties when the work is released and you will get my total dedication to prospecting. You will also have your book written."

My biggest surprise of all: there was no hesitation.

"Deal!" he said. "You won't believe this. My dad and I were thinking the same thing. As a matter of fact, he'll be the man you'll be with. He knows our history, and he knows almost all there is to know about finding base minerals. He is the best prospector alive," he added proudly.

"Wonderful!" I exclaimed. I was very pleased. "Oh, by the way," I said, "you have to call my publisher. It is most important." I hung up the phone.

With that, and with few other preliminaries, I plunged into the prospecting work with Allan Keats, the consummate, always-seeking prospector. Complete with heavy pack, rubber boots, thirty-degree temperatures, four-wheel-drive trucks, four-wheelers, canoes, endless walking, and accompanied by hordes of the finest biting specimens our Newfoundland island has to offer, I would come to love it all. Well, not the flies so much.

I

Soulis Joseph

And a cry came from the people as from a single heart, and it rose the dusk and was carried out over the sea like a great trumpeting.

Kahlil Gibran

1

When the wind breezes high through the canopy and softly swishes down, and the handsome dragonfly flits around you, hunting the bothersome mosquitoes that always seek your blood, it is pleasant to lay on a mattress of damp summer moss beneath the trees, and listen and watch as nature passes along.

Sweat trickled down the channel between my shoulder blades and over my back, seeking the moving parts below my belt. I wiped the top of my scorched bare head and, without removing the heavy bag from my back, paused in my walk to dig a baseball cap out of one of the side pockets. I placed the cap on my head, hiding my scalp from the relentless July sun.

A mixture of salty sweat and fly dope drained into the corner of my left eye and stung like hell. I didn't try to wipe away the misery. I had learned that burning lesson years ago.

Allan Keats and I were trudging through a bed of very dense alders somewhere south of Red Indian Lake in the centre of Newfoundland. Everyone in Newfoundland knows that these common lush growths are to be avoided at all costs. They are not only jungle-like in nature and very difficult to get through; they are also breeding grounds for every insect that likes stagnant water, damp ground, green foliage, and sweating prospectors. They had to be gotten through to reach our destination. The only saving grace for the torment of these tangled beds was the shade they provided, and the damp, cool ground from which they grew.

Al had spent most of last night sitting in silence – unusual for him – before several maps laid out on our cabin table, plotting the course we were now trying to follow. Lines on a map are seldom easy to follow on the ground. Hidden from the paper surface are streams and dense growths which must be gotten through or around or across, while still maintaining the predetermined direction of travel. Even the best of maps, with all of today's technology, never show the underbrush that a person must travel through, most of it "pecker high," a prospector's description of waist-high growth, such as *gowithy*, that I had learned from Al.

"What if the prospector is a woman?" I wanted to know.

"By God, now that's a good point you've raised. I never thought about that before. A lot of female prospectors around, these days. Never used to be, you know. Great to see women doing this kind of work, too. We have to work with one later this summer. When we gets the chance, we'll let her tell us the name she puts on 'pecker bush.'"

Al laughed aloud, his mind racing with the thought.

He announced a stop to consult his GPS and his topographical map. We had just stepped out of the alder bed into a loose mature forest, and faced the direct glare of the sun. He set the navigation device down and waited for the data to come in.

Removing his red pack, Al settled back against a spreading birch tree. He dug into one of the pockets of his bag, produced a bottle of Gatorade, and let the blue liquid run into his mouth while a sucking, bubbly sound came from the vented lid. Not far away I sat against a tree of my own and quenched my thirst with tepid water.

"Don't know how you can drink that bloody stuff," I commented. "You sound like a hungry baby at its mother's breast."

"By God, this contraption *does* look a bit like a nipple, don't it? And me suckin' on it, imagine that." He burst into laughter at the thought, placing the bottle back into the relative coolness of his packsack.

"We're right alongside where we want to be, according to the GPS. Only three hundred metres more, I'd say. The going should be better now that we're clear of them alder beds. Steep, though, by the looks of it. We're coming into sediments. Did you notice that?"

"No, I can't say that I did," I returned.

"You'll learn. It takes time to notice the difference in the rock changes, moreover when the change is gradual," Al said good-naturedly.

"I don't 'low the summer will be long enough for me to learn all the stuff you're trying to pump into my head," I said.

"Well, you said it, I didn't," he replied, his chuckle coming easy.

Another half-hour found us near the top of the ridge we were climbing. We searched for outcroppings and saw several boulders of various sizes. Floats, Al called them. They looked promising, but they could have come from anywhere, "floated along by the glaciers," but not far, he pointed out. They were fairly sharp, not rounded like they would be if the ancient glaciers had brought them from a great distance. It was the location of the contact, where the change was obvious and sure, that we were after.

Al found the first outcropping, of course, identifying the moss-covered mound as a hidden rock ledge. He pawed several inches of moss and tangled roots away with the claw of his hammer, exposing the rock.

"Sediments for sure. Lots of rust showing, carbonated, and even some shearing. Something going on here, without a doubt."

By this time I had walked away from him a bit to find my own outcropping. That didn't stop Al from talking to me. Or was it to himself? I never did find out. We'd be out of sight on a forested ridge and I would always hear his voice in the distance. "Oh, look at the squirrel," or "Luh! God love 'un," or "Well, hello, Mrs. Grouse, where's your family to? Oh, there they are! Luh! God love 'em." I rarely answered his thinking-out-loud on these occasions, and I came to know when his voice was rhetorical and when it was not.

A short distance away from my talking, rock-banging partner, I found another mineral showing. Long since, a moose had climbed a steep mound on its way over the ridge and in doing so had scraped the moss from the rocky ledge, revealing to my "knowing" eye a scrape of rusty rock. The moose had probably tripped, I thought, noticing the break in the animal's stride still evident in the torn topsoil. I was as excited as if I had just discovered gold.

I tore away great strips of the moist green moss and hammered a piece of the rock free from the cliff. It was heavy, and glinted with sulfides and what looked like gold. I blew a sheen of saliva across the rock in my hand, as I had seen Al do to bring out the true colour of the rock. Using the powerful lens strung around my neck, I studied the secrets of the stone.

I saw the gold first. Several flecks of it. Cubed, too. Chalcopyrite – copper – was evident, as well as a greyish mineral. Probably zinc, I figured, as well as peratite. Along the same tree-lined and brush-covered ravine, off to my right, came the sound of Al's hammer striking rock again and again. I turned to shout out my find to him. Then I heard his excited voice.

"Now, would you look at that! Isn't that something, now? All kinds of sulfides. Smells like arsenopyrite, too."

He called the names of several minerals, some of which I had found, including pyrite – fool's gold. But not a mention of gold. He always amazed me with his degree of excitement. It didn't matter what he found: gold traces in a stream, sulfides, rare quartz veins. The simple thrill of finding something that may be of value, and that *he* had found it, was his reward. It was never about money with him, although of course he was intelligent enough to realize the financial significance in any ore body that could be unearthed.

Smell it? That's what he had said. Another sense I had forgotten to use in my evaluation. He had instructed me on "stinky quartz" and even arsenopyrite, from which arsenic was extracted. I had never known, or even thought about, where the deadly poison came from.

I decided to have another look at my own find before showing it to Al. I had shown him so many samples over the last few weeks I was sure were valuable minerals, that turned out to be nothing more than mica, quartz – even rust – and of course pyrite. I had received too many of his chuckles of amusement.

Turning over several of the rocks and selecting the most promising ones, I looked more carefully than ever. I shifted the rock in my hand to get the best possible angle of light.

It could be gold, you know. You can find the stuff anywhere, remember that, even in granite. Hard to tell most times, too. Even I have trouble with it sometimes. It takes practice and patience. The best way I finds is to turn the piece different ways. So the light can get to it, see. Spit on it first. Usually if 'tis gold the colour won't change, no matter how you turn it. Pyrite, now, that's different. It will turn a different shade of its gold-like colour as the light strikes it.

His voice was in my head as if he were standing next to me and quietly explaining the mysteries of minerals.

If it does change, you know, tarnish, like, then 'tis not gold. Mind, though. Pyrite is almost always found in good company – pure gold.

"Pyrite and sulfides over here, too, Al!" I shouted.

"Excellent! Excellent! That's what we came to find!" he shouted back. "We're takin' samples from here for sure. Bring several good ones with you when you come over. It'll be a longer walk back with a pack full of rocks."

Later we bagged several samples in heavy plastic and

labelled them with a black marker. Al marked the coordinates in his book – never on a bag which could end up at an assay company – and loaded it into our packs. It took several attempts on my part to convince Al to load most of our samples into my pack.

"You have a heart condition," I reasoned.

"Yeah, and you got a ruptured disc," he responded.

"Still, my back condition probably won't kill me. Your heart condition could."

"Well, at least I'd die happy, here in these woods that I love, and not only that, doing what I love doing most of all in life, exploring for minerals. You'd do that CPR stuff on me if I had a heart attack, wouldn't you? Don't want to get that image in my mind, though, you with your mouth over mine. Cindy's is the only one I wants there, dying or no." He laughed then, thinking of his wife.

"To tell you the truth, Al, b'y, 'tis an image that floors me, too. Especially with that damn thick grey moustache you're wearing."

We both laughed.

"I've heard you say that before, Al. You know, about dying contented in a place so wonderful as the forest. I couldn't agree with you more. It is a lovely place for woodsmen like us to end our days when the time comes. But what if you don't die?"

"What do you mean?" he asked, pausing in his efforts to hammer the sharp edges of the samples smooth to avoid tearing the bags.

"Well, the way I see it, dying is the easy part of it. You have this terrible pain in your chest, and if it is severe enough, you die, and it's over. But suppose it just badly damages your heart. Or worse, gives you a massive stroke and doesn't kill you at all. There you are, your body twisted and shrunk all on one side.

"Can't talk properly, or maybe not at all. Someone has to feed you . . . nothing, only mashed foods. Do you want to be somewhere looking out a hospital window at the falling leaves and wondering when they'll be along to change your diaper? And all of this with your brain intact?"

"By God," he said. "You've painted a terrible picture. I never thought of it that way, though 'tis one that's all around us to see in any hospital.

"You know what? You've convinced me. I shouldn't be carrying so many of these heavy rocks. We'll fill your pack with all of them, what about that?"

And leaning back to the mossy bank, he expelled his infectious mirth, until we both roared with laughter.

2

Three weeks later, on the Baie Verte Peninsula, our joking about Al's heart condition wasn't funny at all.

The company we were working for, and of which Al was part owner, A.S.K. Prospecting & Guiding Inc., had hired out our services to a large mainland company that had staked mineral claims on the peninsula, as well as on the north side of Green Bay. We checked into the Dorset Country Inn, which we would use as our base in the scenic town of Baie Verte. The place would feel like home for the next two weeks.

Here we met up with Julie Brown, project manager for Takara Resources Inc. out of Toronto. She was a beautiful young woman, slim and soft-spoken, and, we were to learn, very capable. She hadn't earned the M.Sc. and Ph.D. behind her name from classrooms alone. She also had some Newfoundland roots.

With Julie was a young geology student. Daniel Weidendorfer was from Germany, and as part of his training he was doing fieldwork in Canada with Takara Resources. He was overjoyed at having been sent to Newfoundland. Al and I immediately christened him Danny B'y.

One evening, another addition to our party arrived from the Deer Lake airport: Scott Jobin-Bevans, Ph.D., P. Geo., Vice-president in charge of exploration for Takara; First Vice-president of the Prospectors and Developers Association of Canada (or PDAC); and Managing Director for the Americas, Europe, and Asia for Caracle Creek International Consulting Inc. Julie had already given us his cards.

Just great, I thought. A week spent in sweltering heat with an over-educated, fancy-talking executive.

He came through the hotel doors with an outstretched hand as soon as he saw Al and me. A full head of curly blond hair above a smiling, easygoing face topped his six-foot frame.

"'Ow ya gettin' on, b'ys? I'm Scott." He greeted us both with a strong handshake. Nothing stuffy about this guy. Our little expedition was ready.

For the next couple of days we roamed the hills near and far, leaving the hotel early each morning in two trucks. The temperature held at thirty degrees. Fortunately a brisk south-west wind rose with the sun most days and helped cool us. Al and I did the farthest exploring, walking for miles over boulder-strewn ridges and rocky barrens that stretched as far as we could see. The others investigated their claims closer to the woods roads.

Daniel had never seen a moose or black bear. Each evening when we met back at the hotel we would tell him of moose we had seen on our travels. We also told him that the only way he was going to see moose was to come with us. Soon we were all good friends and over supper shared the day's events and findings while planning the next day's trek. Daniel struggled with some of the language, especially with the way Al and I talked.

"You talk so fast," he said, to which Al replied, "That's not it a'tall, Danny B'y, you're just listenin' too slow," and laughed in his good-natured way.

Scott was in constant communication with his mainland offices. His evenings were consumed between his laptop and the phone. Only two mornings after his arrival he greeted us in the lobby with his bags packed and ready to go. He had to fly to Winnipeg that evening.

"We'll do that traverse we planned last night," he said. "I'll leave from the field and drive straight to the airport." He had his own rental truck.

After we got in over an old, logged-out road as far as possible, everyone piled into our big truck. It was equipped with four-wheel drive. We would need it to get over the deserted logging road, which was now used extensively by ATVs. It wasn't wide enough for our truck, but Al drove on anyway, relieving most of the encroaching alders of their outer limbs and most of their green foliage, which fell into the back of our truck as we passed.

We were looking for uranium. I carried a scintillometer, which gave off a series of high-pitched clicks that increased when coming into contact with radioactivity beneath the earth's surface. It cost around twenty thousand dollars. I was told to keep it from getting wet at all costs. I figured for that kind of money it should come equipped with its own personal flotation device. It didn't.

Uranium is found throughout most of the world in low-grade deposits, but Canada is one of the few places where high-grade concentrates can be found. Daniel was proud to tell us the element uranium was discovered by Martin Klaproth, a German, in 1789.

We climbed a rocky ridge that had seen a major forest fire a few years earlier. It was stripped of all but new, low growths of green shrubs and trees which struggled for life between countless naked, white boulders. Every ravine and deep gulch was filled with the tangled blowdowns of a fire-ravaged forest. The going was very difficult.

Al kept up his pace despite the terrain. He was well ahead, and I was in my preferred place, in the rear. Scott asked me if Al always walked this way. I assured him he did. Even faster on good ground.

"Hey, Al," he called up the slope. "Slow down, b'y, the rocks will keep!"

Al laughed as always, waited for us to catch up, then set off again.

Blueberries were ripening in clusters from every available patch of black soil. Stunted pin cherry trees, young aspen, and white birches flourished. Nature was already at work, growing first the deciduous trees that would shed their leaves every autumn and start to rebuild the soil. The young growth was greening the drab landscape, which looked like it had been the scene of some fierce battle.

We found samples of pitchblende ore in the sedimentary rock that Scott's geology maps had showed him the night before. The counter gave off an almost constant static as we moved along, and in one spot accelerated its noise, its dial showing high readings. It was very promising. We took several samples along the barren ridge. The air up here was stifling; the slope faced away from the faint breeze.

Julie and Daniel had reached the summit of the hill and were staring at the burned-over country that stretched for miles in either direction. A sudden movement from his peripheral vision drew Daniel's attention to his right. He was so amazed he barely managed to whisper Julie's name. With his thick German accent he always made it sound like *Chulie*.

There in front of them both and in plain, easy view was a large bull moose. Daniel hurried his digital camera from his pocket and raised it into position, only to see, full in its lens, not only the moose, but a black bear as well! His jaw dropped, but he remembered to snap away with the camera, capturing forever the rare image of two of our island's largest animals. His day, and entire trip, was complete.

Later that day we bagged our samples – double-bagged, and sealed – under the watchful eye of Julie. She would take no chances with radiation, no matter how small the levels. We carried them on sweating backs down the hills to the truck.

That evening found us back near the highway. Scott was running late for his flight. He would have to hustle, he said. Obviously this wasn't his first time making do. He stripped to the waist, washed in a small brook, and hid behind the truck to change his trousers and underwear. After stuffing his clothes into a well-used duffle bag he drew a comb through his matted hair, shook hands with us all, and was gone.

The next morning the four of us gathered under the same ridge preparing for an even farther and higher climb. The August sun had not relented. The temperature hovered near thirty degrees.

Before we were long on the trail, I removed my shirt, as I always did on warm days. I glanced ahead at Al's white head bobbing up and down as he picked the best way for us through the snarled, uneven area.

I had taken my eyes from him and was looking to my own footing when I glanced back up the trail at the others. There was something wrong with Al. I kept looking, as much as the terrain would allow, but still couldn't absorb what my senses were trying to tell me.

I stopped and stared after our plodding leader. And then I knew what it was that had caught my attention. Al never plodded anywhere. There was a definite break in his stride. Something was very wrong with him. He was tired, but he wasn't about to let anyone know. I was sure of it. Still, it was a very hot day and our way was steep and hard. Maybe I was wrong. I kept watching.

I had walked behind this man through heat and rain, up hard hills and down, across muscle-draining bogs and through the heaviest of woodlands. The only time he ever slowed was when we reached our desired area and would begin our search. We were far from that area yet.

His hand went out to a large boulder as he passed, as if to steady himself. He stumbled, once, unnoticed by the others.

"Al, b'y, we'd better take a break. I'm beat, ol' man," I called ahead.

"Oh, all right, b'y, we can take a spell, you knows we can. Not a problem," he called back and stopped at the nearest boulder. He was the first to sit down, even before I had reached him. Another uncharacteristic display for him.

He was breathing hard when I walked up to him.

"What's wrong, Al?" I asked in a tone that said *Don't bother telling me nothing's wrong*.

When he glanced up at me his brown eyes were sad. Tired.

"I – I don't know, b'y, but I – don't feel right. There's a kind of weariness in my chest. I'll take a shot of nitro and then I'll be all right," he said. "Don't tell the others," he was quick to add.

It was too late for that. Julie had sensed the seriousness of our conversation. Both she and Daniel walked up to us.

To their concerned query he replied, "Must be a bit of angina that's come against me, sudden-like. Don't worry. I'll take this medicine and I'll be just fine in a few minutes."

"Why don't you slow down, Al?" I asked him. "Scott was right yesterday. Those bloody hills aren't going anywhere. Besides, this heat is taking a toll on us all. Not to mention the three stents you have in your heart. Add to that a mountain with terrain even a fox would walk around, and a smart feller like you should see the need for restraint."

"You're right. I know," he replied, his chuckle almost back to normal. He drank water then and rested. We told him we weren't moving for at least thirty minutes.

Julie and I decided – out of earshot – that we would finish the traverse alone. Daniel would walk back to the truck with Al. If he wasn't feeling better in the next few minutes we would all go back – or call for a chopper if he was getting worse.

"Nitro working yet?" I wondered aloud to Al.

"Yes, b'y, I feels a nice bit better already. Good thing, too. My doctor told me that if I have to spray the stuff a second time, I'd better be on the way to the hospital. I s'pose that would have to be a chopper ride from here, eh?"

I agreed, and looking out over the land, I felt that though this place was very inhospitable for hiking, it had a beauty worth coming to see all the same.

We reached the crest of one of the hills that was part of our travel plan. We were high enough to receive the full sweep of the summer breeze. It was a soft, scented wind that brushed over the land beyond and picked up the smells of ripening berries and the warmth of the earth. From up here I could see in the distance where the ocean reached in through the green land far below and shimmered in its summer finery. The clouds and their shadows raced across the valley below our gallery seat. The view was spectacular. Intoxicating.

The medicine worked like magic, and in just a few minutes Al was feeling much better. We told him of our plan to send him back to the truck. He wouldn't hear of it at first, as I expected.

"We can't let you go back alone, Al, even if you are much better. You know that. And we can't send Julie and Daniel on alone either. So if you are feeling okay, Danny B'y will go back with you and I'll go with Julie back down the other side of this ridge, as we planned."

Daniel told Al he agreed with us. He would gladly give up the trip and go back with him.

He still wanted to go back alone. He'd be all right.

Julie said no. She was responsible for our little excursion, and she would feel much better if Daniel accompanied him back to the truck. If he didn't agree to these conditions, she would scrap the exploration.

"No, no, my love, there's no need for that. I'm fine now, just overdid it a bit is all. I'll be all right. No need to cancel the traverse. My buddy will see you all the way over, don't worry about that. Danny B'y and I will walk back to the truck and drive in from the other side. That makes more sense anyway. Not so much walking involved."

Soon after assuring us that he was fine, and after Julie was convinced that he was good as new – and getting a promise from Al that he would take his time – they turned back down the hill. They hadn't gone far before Al shouted back at us.

"Don't worry, there's lots of rocks left for me to bang yet!" His chuckle was back where it belonged.

Three hours later, without finding anything of great interest among the mountains of rock, Julie and I descended the hill. It wasn't easy going, to say the least. There was no trail. I told Julie the locals had too much sense to try and walk down the way we were going. There was probably a well-worn ATV trail nearby that we didn't know about.

After stumbling through the net of dead, fallen trees and scaling at least two cliff faces, we reached the brown bog below, ending our traverse just beyond the painted fences of an unsuspecting town. Not once did Julie complain about the terrible hike down over the mountain.

We called the truck. Daniel answered and assured us Al was his old self. Thirty minutes later we were riding along the highway, headed for our hotel.

Two days after that we said goodbye to our two young friends and left for another adventure, somewhere else. We never did ask Julie about the "pecker bush."

3

Anchored like a huge jagged triangle to the cold heart of the North Atlantic Ocean, the island of Newfoundland juts into the path of two of the world's mightiest ocean rivers. The Gulf Stream, flowing north from the heat of Mexico and the spew of the Mississippi River, shears away from the southern part of the island, while pouring endlessly from the frigid north, the cold Labrador Current bends around the island's north coast. Both currents slide away from the east face of Newfoundland to the expanse of open sea, as if not knowing how to approach this brazen bastion of rock that diverts their salted flow.

Both ocean rivers leave behind great mysteries of life and untold volumes of living protein. Here, where the cold north collides with the warmth of the deep south, the merging of waters spawns a sea life unlike any other on earth.

Lying just inside these ecological wonders, The Rock awaits, as it has always done. From the distant ocean this land indeed lives up to its nickname very well, towering over savage seas feared by all seamen. The approaching island brooks no entry until the newcomer is about to veer away. Then the cliffs open up to reveal a sweet-smelling land whose green verge reaches down to the very sea edge to welcome all sailors.

Long before João Vaz Corte-Real of Portugal made his late "discovery" of 1472, other Europeans had come here, their brief stays only now being unearthed. Their intrusion on this green isle eradicated an extraordinary, red-painted native people, whose very existence is recorded on the smallest of pages.

Whoever its discoverers, ancient and new, The Rock it was and is. It still awaits further discoveries. Its strata is unchanged, for the most part. The very name "Rock" implants in the mind a sense of security and stability, a steadfastness, but this constant island home, awash in a sea of change that creeps inland from our jagged coastline, forces its present-day inhabitants along a road of inevitable collision with the rest of the energy-hungry world.

Today's modern explorers inquire after the treasures hidden beneath the hard surface of this rock, where great secrets are guarded by some of the harshest terrain on earth. Minerals and oils await human discovery. The breed of man that searches for these treasures is unknown to most of us. Their way of life is a mystery, but their work goes on all around us, unsung and unseen, despite our appetite for their hard-won discoveries.

Allan Keats and I followed their indelible footprints all along the western end of the golden shoreline of Red Indian Lake. I could feel their presence watching from the shadowy woods.

We waded across a stream that tumbled down from the hills and wound like a short black ribbon across the landwash before losing itself in the depths of the huge Red Indian Lake. I was flushed with a sudden feeling of intrusion, as if I had no business on these sacred shores. This was the home of the red-skinned people whose name lives only in scattered, incomplete history pages and on brochures of brief interest in a few tourist chalets. Their very existence has been erased by men of my ilk and pale colour. I shuddered at the terrible emotion that seized me, as if I were somehow to blame for the corruption levied against a vanquished people by my marauding ancestors.

I have always been fascinated by the Beothuk. They had been the true Red Men, their bodies stained by the crushed red ochre mineral found on this island.

Columbus and other Europeans were right in theorizing that by sailing west they would find the riches of India. They were of course mistaken in thinking that these huge North American lands were the gateway to the Orient. When their heaving caravels found shelter from the rolling sea in one of our countless bays, and smelly, bearded men stepped ashore onto pristine soil, they were watched by a people who lorded over a land that held unimaginable riches. These virgin treasures would be the undoing of an entire race of a proud yet private people. Though these land-hungry invaders from Europe would have to sail around another huge land mass and across a different ocean, and still be thousands of miles from India, they were convinced they had reached their goal.

Glimpses of "Indians" were cause for great concern among these discoverers. Not only did they call them Indians – thinking they had reached the Indies – but Red Indians, because of the colour of their skin. The words "Red Indian" stuck, and would be synonymous with all other natives discovered farther along the vast western continent to which this angular island pointed.

* * *

We sat for a while enjoying the misty coolness coming from the huge expanse of water. It was a very hot summer's day, the temperature reaching thirty-three degrees. There was no wind at all. I told Al about my eerie feelings for the lost tribe of our island.

"They're still here, you know," Al said, sipping his blue Gatorade.

"How do you mean?" I demanded, gulping the last of my bottled water and looking into the woods, half expecting to see a Beothuk Indian peering out.

"Well, you knows, o' man. Those people had their pow-wows. Well, that's the same as parties to us now. Point being, in some cases over all them years, them two tribes had to come together, you know, the Micmacs and the Beothuks. They had to. Fishing for salmon along those rivers, huntin' the caribou over the same areas. It don't make sense that there wasn't some contact. Friendly contact, too, I mean.

"And on a beach on a warm summer night like this night is going to be, them young Indians snuck away and had sex. You knows they did. 'Tis only human nature, b'y. All they have to do now is find the DNA of the Beothuks and they'll find they're still among us. I'm convinced of it. Wouldn't that be something?"

Al had become excited by his thoughts. His dark eyes flashed, relishing the idea of ancient Red Indian blood still coursing through someone's veins. Especially his own. I told my companion that I agreed with him. I knew of many others who shared the same opinion, with their theories passed down through the years from the old ones, of how they were direct descendants of the Beothuk.

We talked some more about the native peoples, how they must have suffered to survive, how clever they were, what an amazing paradise they possessed, and how very sad it was they were gone.

"Can you imagine hunting caribou with a bow and arrow, or maybe a wooden spear?" Al asked me. "And when you killed one, havin' to clean it with a crude knife made from a piece of slate rock or chert. They were some smart people, b'y. I only feels like half a man when I compares myself to them."

We talked some more about the old Indians, both Beothuk and Mi'kmaq. My own feeling was that these first inhabitants were quite contented here. With no comparison, this secret island was their home, in which survival depended on their own ability and wit. Not much has changed, only the names and faces of the people. The struggle still continues.

Our talk slowed to a few murmured comments as the heat of the noonday sun crept over us. The blue lake water lapped against the shoreline, while the warm, fragrant breeze tickled the summer green that rose to great heights behind us. Lying in pure and peaceful comfort and with nature's best lullaby orchestrated for us alone, Al and I closed our eyes and kept all further thoughts inside.

4

Two British sailors strolled along the rickety wharf in the busy harbour of Halifax. Night was coming and the late evening became cold very fast with the sun's leaving. It was late October; twilight would not linger long. Hardwood trees all around the harbour were in their full autumn splendour. Their colours shook and swayed on the breeze coming from the nearby ocean. It was very beautiful. The two sailors never noticed.

Here along the shoreline were smells that didn't blend well with the age-old forest scents: coils of cordage, of manila and hemp ropes, as well as looped skeins of deep-scented oakum, and empty tar buckets. On every dock, overfilled barrels of rancid cod-liver oil dripped and leaked and greased the wood surfaces.

Two ships clung to the structure, their sails furled with loose lanyards. As the tide pulled they stretched their lines taut in a vain effort to be away to the clean ocean. One of the vessels tied to the wharf was a schooner laden with quintals of salted codfish waiting to be off-loaded. The other, which the two sailors were headed for, was being loaded with new-sawn lumber. She was near-loaded with white pine and fir timbers. A third ship, a brig, was anchored away from the dock and swinging on her mooring midstream.

Several outhouses stuck out from the rocky landwash. The tide was out, the smell of their human issue not yet flushed away by the evening flood. But overpowering it all was another smell, a stench so vile and nauseating that it caught in the throats and watered the eyes of the two seamen as they walked along.

"Wot the bloody 'ell is that smell?" the taller of the two exclaimed. He hawked from his throat and spat a stream of black tobacco juice onto the top of a grump which held one of the ships to the land. His partner retched a bit as well, his open mouth showing a few rotten teeth browned from constant chewing of the cured leaf.

"By gawd, 'tis a smell I won't forget, though I niver thought I'd be sniffing the foul breath of 'er along these shores."

"Wot are ye talking about, Fletch? The foul breath of who?" his friend interrupted, looking around for the "her" to which his companion had referred. "There's no woman about that I kin see."

"Not 'er as in a wench, man," Fletch replied. "'Er wot's lying moored midstream there, with her dirty arse pointed our way and the evenin' breeze sendin' her black perfume our way. 'Tis a blackbird, by gawd."

"Black wot did ye say? Did ye mean that brig yonder is where that gawdawful smell is comin' from?"

"Damn ye fer the dumb Jack tar that ye are. She's a blackbird, I tell ye, a Guineaman, a slaver, fer gawd's sake. The curse of all good sailors, a blight on our noble trade. An honest seaman fears 'er as much as the press gang. She won't come alongside the dock. 'Shamed, she is. She swings on 'er chain in midstream like the dog that she is. An' she flyin' the Jack proud as ye please from her topmas'," railed Fletch, his anger at the sight of the slave ship so close suddenly animating him, his arms and hands shaking and pointing at the hated ship.

"Do ye mean there be black nigger slaves aboard that brig now?" asked the other seaman. He seemed excited at the prospect of such an event.

"No, man, ye'll not be finding the poor devils chained below 'er decks this far north, though I 'ear they be brung 'ere by other routes overland. No, this one is clean of 'er niggers right nuff, but she can't rid 'erself of 'er stench, kin she? No amount of sea flushin' kin cleanse 'er. Bad luck to 'er an' all blackbirds."

His mate tried to interrupt. He had many questions to ask, but Fletch ignored him.

"If ye sleeps yer grog off in the alehouses of Lon'ontown, make sure you've a buddy to keep yer watch. A billy stick across yer noggin will land ye aboard one of these wretched ships, whar ye'll be treated little better than the sorry creatures below her stinkin' decks. 'Tis better to be pressed into the King's navy, I tell ye. Aye, even the King's navy would be better than sailin' on the likes of 'er."

Fletch was visibly upset at the sight of the slaver now exposing her broadside as she leaned against her hook.

"The only good thing that comes from her 'old is the

sweet spiced rum that she brings from the Indy Isles, an' only then if swabs like us kin get our hands on a bottle before the British merchants get their greedy paws on 'em and waters 'em down. Waters all of the spices out of it, they do. By gawd, 'tis a throat full of stingin' rum, spiced or no, I'm needin' now after sech a sight as that one. Though 'tis no sweet Indy rum we'll be gettin' this night."

And with a hurried walk, Fletch headed away from their own ship tied alongside, fumbling through his deep woollen pockets in search of rum coppers as he went.

His fellow seaman was still standing spellbound, staring at the dark ship of legends. Her putrid smell filled his senses. He turned to tell Fletch he was right about the stench, that he would never get it out of his mind either, and only then realized he was alone on the ill-kept wharf. The shadow of the slave ship fell across him as the night deepened. He felt an awful flush steal across his upper body; he was sure he was being watched, though he saw no one. His legs felt weak. He swallowed back a thin hot bile that crawled into the back of his throat, and with a weak yell he raced after his disappearing mate.

* * *

Two mornings after, as the early grey dawn brought light into the harbour, the brig manoeuvred alongside the wharf. The other British ship, loaded above decks and below, slid away on the pre-dawn tide, her crew staring over the bulwarks in total defiant silence at the empty slaver docking in their empty berth.

Soon the high-pitched squeal of a rusty windlass rang out across the morning harbour. The schooner with the load of salt fish in her hold bustled with working men. Wooden tubs filled to the brim with the salt- and sun-cured codfish were swung up out of her black hold and lowered onto the dock. Men hoisted the well-used tubs a few inches above the wharf decking while the stern British captain weighed them. Then they loaded the containers high onto wooden hand barrows and carried them up the slaver's gangway.

Codfish was a staple for the British merchants. This particular load came from the south shore of Newfoundland. The best of it was sold at high prices to the general public. Hundreds of pounds were purchased at exorbitant prices by the British military. Cod which had been pulled from the waters of the Western Atlantic were found to be an excellent and plentiful food source for Britain's army and navy. In a sense the small island nation of Newfoundland ruled the greatest empire in history, which was so expansive that the sun never set on its flag.

The salt fish was cooked, without first watering some of the salt out of it, and given to marching footmen and storm-tossed seamen alike. Even when the fish spawned maggots and smelled rancid, it still had to be eaten. Profits mustn't be lost. This spoiled and decaying fish often found its way into the holds of the outbound slavers, next to the imprisoned Africans who lay and tried to sleep in spite of the smell.

Below the thick decks of the brig, three men had been given the back-breaking task of stowing away the hundreds of pounds of rotten fish that came aboard. From above the decks and through the only light-giving hole, more men flung the cod below to the struggling, cursing trio. The bulging hold in the waist of the ship was already packed with great wooden barrels of rum, vats of molasses, boxes of spices and raw sugar, as well as huge tied bales of raw white cotton purchased from the plantations of Virginia. Despite the different smells the casks and barrels and bundles emitted, the overpowering stench of the brig's recent cargo so dominated the dark hold that even the pungent fish could not overpower it. The brig's wooden frame would forever be perfumed with it.

One of the men below decks had been given the job of stowing some of the fish that was thrown his way into the few cramped recesses between the cargo. Twice during a brief lull in the work he thought he heard a sound coming from the black depths of the ship, though straining his neck and peering as hard as he might he saw nothing. He was filled with fear. The smell stuck to his throat and scoured his nostrils. He had never experienced such a feeling before. It was a trapped feeling, of despair and hopelessness.

An angry shout from above halted the work for a while. A line to the windlass had broken, spilling a tub's contents back into the hold of the schooner anchored near the brig. The two men directly below the skylit hold took a grateful pause in their labour. One dug in his pocket and produced an unwrapped plug of tobacco. He tore off a large, jagged piece between his teeth with several shakes of his head. The other turned his back and pissed, most of the urine falling on the stacked fish.

Farther back and in the shadows, the other worker was about to sit for a much-needed rest when he heard the sound again. No mistaking it this time. Something or someone was back there. It was a shuffling, stirring sound. But more than that, he was sure he had also heard the clinking sound of chain this time. He waited. From the same direction came a low moan of pain and misery. He thought maybe the captain had a dog tied up back there, but there had been no barking, nor any whining as would come from a dog kept hidden in such a manner.

He glanced back toward the light and the other men. They weren't paying any attention to him, nor would they. They almost never spoke to him, except to curse orders at him.

Standing erect, he found a passage of sorts through the stored goods. It twisted around barrels and wound over boxes as he followed the sound. Deeper into the cluttered, tomb-like hold he continued, until he felt he had gone far enough. He paused in his search. For a moment he thought he was mistaken, for now in the silence of the brig's bilge he heard nothing.

The smell here was putrid. He felt a presence, very close. He heard breathing, and then, in the same instant, he was startled by a pair of white eyes that stared at him without blinking. He still thought it could be a dog. But now, as his eyes adjusted to the darkness, he could see it wasn't a dog at all, but a man unlike any other he had ever seen. He wasn't a white man, of that he was certain. The only thing white on him were his eyes, which still hadn't moved. The watcher shifted from his crouched position and allowed a little more light into the space where the chained man lay. He was chained, there was no doubt. The prisoner was lying on his side, his knees bent in a fetal position. He couldn't straighten even if he wanted to. The ledge he was lying on wasn't long enough for him to stretch his length, nor was his perch high enough to allow him to sit up straight. All of this the watcher saw in a glance, and more.

The prisoner's skin was black. The stark white of his eyes only added to the inky blackness of his skin. He was barely clothed; what little he wore was torn and shredded, showing black flesh. Just below one knee his pants had torn away, so that his leg was naked. His other leg fared little better. He smelled terrible. Just below the big staring eyes, thin grey streaks of skin showed, as if tears had run out of his eyes in an effort to cleanse his face from the filth that covered it.

The worker stared in fascination at the first *antaqvej* – black man – he had ever seen. Even his hair showed no contrast to the black skin. Unruly dense black curls of it draped the once proud head and fell over his forehead, so that if he lowered his head his face would be hidden. Below his broad chin a dark red scar, about two inches wide and not yet healed, disappeared beneath the hanging black locks. Both of his wrists were collared with bands of iron that were polished from constant use. The silent watcher noticed the iron bands were the same width as the scar on the black man's neck. Two more manacles were fastened around his bony ankles. Secured to the four iron rings were two pieces of chain that appeared to be strong enough to hold a draft horse. The chains ran from the slave's limbs and were clasped to the beams of the ship. There would be no escape. Where the iron met flesh, pink skin showed in different stages of healing and breaking open again. The prisoner had endured much misery and for a very long time. He appeared to be young. About his own age, the watcher decided.

The black man had once been strong and heavily muscled, his loss of weight evident from the slack in his iron shackles. He blinked, or rather closed his eyes – the whites of them disappeared for several seconds – as if when they opened again he would be free from his torment. He moved slightly to ease his pain and the same low moan the watcher had heard before escaped his mouth.

In all this time not a word was spoken by either of the two men. It was as if both understood talk would be useless. Beside the black man a dirty, overturned wooden bowl lay empty. The watcher reached into his pocket and found several pieces of dried meat. He reached out his hand and offered it to the slave, who just as slowly took the food. For the briefest of seconds their fingers met, and with that small contact of human compassion, a flicker of hope ignited in the eyes of the black man.

All of a sudden, from outside came a harsh curse, as the other workers yelled for the watcher. He never understood the curse words they yelled at him. His own language had no curse words at all. He bent to crawl toward the light but stopped as the chain behind him clinked once more. He looked back and found the white eyes staring at him as before, and in that instant he knew where he had seen that look before.

He had once found a lynx trapped by both legs and still alive in its snare, several miles along his trapline. It had been raining for several days. The animal's rich fur was dishevelled and plastered to its spare frame, its tufted ears flattened along its small skull, and both legs were bleeding and torn where the snare had bitten. The trapper knew it had suffered long;

the animal's agony was plain in the unblinking angular eyes that stared at him as he approached with stick raised.

Another angry curse followed the bark of his name, and Soulis Joseph, an indentured slave himself, crawled away from the snared human and back into the white man's world.

5

I came awake as easily as from a gentle sleep on a soft bed and for a second stared at my surroundings, expecting to be somewhere else. Al was lying awake and staring skyward with his hands folded underneath his head. He looked at me without his usual smile.

"Thinking about them old Indians, wudn't ya." It wasn't a question.

"Something like that," I replied, wondering how he knew. My voice was hoarse as I tried to separate my head from the vivid images that had claimed my brief sleep.

* * *

They had come – Al's ancestors, the Mi'kmaq – from Nova Scotia, or so the historians tell us, and there is evidence that shows they were here during the Beothuk's time, at least on a seasonal basis. The first of these people, like so many others of the world's indigenous tribes, were explorers.

The oral history of the Mi'kmaq people says they have

been here since ancient times. Long before the whites ventured across the Great Water. There are also recorded references, both French and English, of whole families of the Mi'kmaq people travelling back and forth from Cape Breton to Newfoundland's south coast, and even as far as Placentia Bay. Why are we to believe that before the Europeans came, the Mi'kmaq just stared across at a new land and never attempted to reach it?

The brown-skinned native Phoenicians were probably the first people to find their way out of the long blue bowl of the Mediterranean and dare their way out over the mysterious, eastern Atlantic in their "canoes." These people were seagoing merchants at a time when the world was being explored and razed by marauders and their endless wars. They crossed the wide Mediterranean and voyaged beyond. They traded cedar timbers to King Solomon for oil and wine. These people found their way south into the Indian Ocean and as far north as Caledonia, long before the Romans had given their name to the top of the British Isles, and before the white-skinned men of Europe had ventured beyond the ocean horizon at their doorstep.

The Polynesians, who left the Malaysian peninsula and explored the South Pacific all over, sailed as far as four thousand miles from their island homes. Out of sight of home they carried the frigate bird with them. This bird would not settle on water, knowing the sea would rob it of the ability to take flight. When released the bird would follow its instincts and fly toward the nearest land, and the eager explorers would paddle after it.

Some have no trouble believing these fascinating stories,

but for some reason doubt that the Mi'kmaq natives of our own land could sail or paddle the few miles from mainland Canada to the island of Newfoundland. The island the explorer John Cabot had called St. Paul, never knowing or caring that it already had its own native name, is situated fifteen miles northeast from the fingerlike tip of Cape North on Cape Breton Island. Here the North American continent ends its efforts to connect with the big island of Newfoundland, just forty-four miles to the northeast.

The Mi'kmaq people called the island of St. Paul "Kewtuminege," or Round Island. These hunter-gatherers used this island for centuries. It was inhabited by thousands of seabirds, which provided not only meat but thousands of eggs in season. It even had a couple of freshwater ponds. The stunted growth of coniferous forest on the island provided a measure of shelter for a very hardy people. Its highest central hill, almost five hundred feet above the sea, provided a far-seeing view of the ocean to an inquisitive people.

From the craggy, white-laced shores of St. Paul Island the flat tops of the Long Range Mountains, etching the eastern horizon and rising out of the ocean like great mesas, end with a plunge into the Atlantic Ocean at Cape Ray, Newfoundland's southernmost reach toward the huge land mass of North America.

Viewed from its highest point, the island of St. Paul looks like a grotesque, full-bellied sea serpent. It is swept and ravaged by the might of the Gulf of St. Lawrence's tide and winds, twin terrors in the days of sailing ships. It has earned the nickname "Graveyard of the Gulf" honestly; its sparse topsoil holds hundreds of bodies, some of them in mass graves, all victims of a treacherous island rising out of a troubled sea.

From the top of Kewtuminege, looking toward the rising sun, adventurous young Mi'kmaq men stared long out to sea. The native Canadians the white men had wrongly called Indians were constant searchers, ever curious and restless.

Soft summer winds from the southwest dominated the lengthening days. A distant land beckoned, almost invisible on such days, and lying low on the ocean just under the sun. The canoes they used to get to St. Paul Island were longer and wider than the ones they used in the rivers and lakes. Still made from layers of the same tough, one-piece bark of large white birch trees, they sported high triangular sides at mid-beam that made them stronger and more seaworthy. The first daring explorers who paddled out over that virgin sea in their fragile barks were as brave as Columbus or Cortez, or the most daring of all the adventurers, the Vikings.

The Mi'kmaq people believed in several gods, but one supreme spirit. Maybe they were drawn to the huge island by spirits not unlike Homer's ancient Greeks, who were tempted by the songs of sirens on the watery edge of their world. On clear, pleasant summer days the bravest of the Mi'kmaq found this edge of their known land was just a stepping stone to another wondrous land that held untold riches.

But the huge cliffs and elevated land of western Newfoundland, viewed from the shallow draft of a leaky canoe, must have been a daunting sight. Here the great maw of the Gulf spewed out and sucked back its everyday issue of swollen tides, and these weary seafarers had only a southwest wind and a following sea to help them along. Great flat-topped mountains like none they had ever seen lined the closing horizon. Sheer, chalk-white cliffs offered no entry, no welcome landing for the natives. Then, just as hope, like the tide, was at its lowest ebb, they saw, through the spindrift, *noywa'mkisk* – where the sand blows.

A beautiful sandy spit reached out to meet them and led to a peaceful sheltering bay. The long stretch of water that the white man would name for yet another of his patron saints – George – would lose its original native name, but never lose the historic footprints of one of its first native peoples, the Mi'kmaq.

It would take the Mi'kmaq many more years to circumnavigate this indented land fraught with untold dangers and mysteries, as well as great treasures in store for these fittest of peoples. Their constant exploring and marine wanderings would prove they were on an island – a massive one, for sure, and with countless bays, all of them different, but an island just the same.

Without even knowing it their lives would be determined by political events they never understood. France and England had been waging war with each other since before the Crusades, when pious English monarchs led ill-equipped armies across the European nation and on into Asia, pillaging, stealing, and raping as they went. All in the name of God, glory, and riches. Their efforts, after tramping across Europe and on into the Middle East to rescue Jerusalem from the infidels, did nothing less than keep the plains of Central Europe in almost constant battle.

The breadth and roll of the North Atlantic could do little

more than keep these two mortal enemies at bay. Even here, in this western world not yet polluted by religion, the hand of the white man's God would play a major role in the future lives of almost all of the native peoples. Again France and England would fight. Kings and church officials would determine the lives of people and entire nations they had never seen, and again innocent races would be drawn into a war that was not their own.

Although it stalled the conflict between the warring nations of Western Europe, the Treaty of Utrecht would have a devastating effect on the lives of native people in Canada. It would make legal the buying and selling of another race of people, changing the course of history, and would determine the outcome of world politics to this very day. In 1713, in the city of Utrecht in the Netherlands, the deal was made, the parchment was unfurled and duly signed, and British ships sent away to the high seas long before black ink had dried on yellow parchment.

The trade had gone on for more than a century before. But now the powers of Europe had made it legal to deal in flesh. The Spanish gave to the British *asiento*, a slave-trading monopoly to supply to the Spanish colonies in the Americas, male and female slaves from Africa, young but not old.

Incredibly, by signing the Treaty of Utrecht, the French nation also ceded to the British all of the land owned by the Mi'kmaq peoples in all of the New World. This would include a land base owned by other native peoples stretching all the way from Hudson's Bay to Newfoundland. The French held the right to continue to fish from Point Riche on the west coast of Newfoundland and as far south as Cape Bonavista on the east side. They also held onto the islands of St. Pierre and Miquelon, which they would never relinquish.

They kept and colonized the Isle Royal – Cape Breton – as well as Isle Saint Jean – Prince Edward Island – and founded Louisbourg, where they built a modern fortress. All of it thrived on the riches derived from the codfish.

The native Indian involvement in this deal, which would forever affect their lives, was non-existent. Almost a full year had passed after the signing when Governor T. Caulfield excused the Mi'kmaq from "a breach of peace" upon a British vessel, because "They did not know that the treaty was concluded between the two crowns, or that they were included in it."

It wasn't until 1715, two years after the document was sealed, the Mi'kmaq nation were informed of their terrible plight. Two low-ranking British officers held an informal meeting with a few of the native chiefs and informed them that all of their lands, as well as the control of all of their people, was now transferred to King George of England. They were further told that King George was not only their new king but their great father across the sea.

The Mi'kmaq chiefs informed the two officers that they didn't know, nor would ever recognize, a king that they had never seen from over the Great Water. The chiefs pleaded their case with much sign language and broken English. The English must be mistaken – the native people had no word for "lying" – they had not given over their lands to the French king. They were never his subjects. How could the king of France give to the English king what wasn't his to give?

Their pitiful arguments and emotional pleas for justice fell on deaf ears. The Mi'kmaq of Nova Scotia tried again and again to have their lands restored, as well as their right to live their own way. In desperation they directed their protests to St. Ovide De Brouillant, governor of Louisbourg, who responded this way: "The French king knew full well that the lands on which you tread, you possess them for all time. The king of France, your father, never had the intention of taking them from you, but had ceded only his rights, to the British crown."

This also fell on deaf British ears. The terrible deed was done. Britannia once more had flexed her muscles.

Over the next few years, the Mi'kmaq people tried to retaliate against this injustice and terrible indignity. There were skirmishes along the coastal towns, some of them resulting in death on both sides.

Their pleas to the English king – through his lace-collared governors – for the return of their lands amounted to nothing. This Indian nation would forever fight, and even beg, to have their stolen land returned to them. It was all in vain.

By October 1, 1749, the governor and founder of Halifax, Edward Cornwallis, had had enough of their useless prattle and interference with his authority. He ordered: "That a reward of 10 guineas be granted for every Indian Mi'kmaq taken or killed. That a company of 50 men be raised to scour the woods around Halifax. That a company of 100 men be raised to go over the whole province" and "to annoy, distress, take and destroy the savages called the Mi'kmaq Indians. The promised reward to be paid upon producing such savage or his scalp."

The bounty was to be paid for the scalps of men, women, and children alike, regardless of age.

On June 21, 1750, dissatisfied with the number of bloody scalps taken, Cornwallis raised the price on Mi'kmaq heads to fifty pounds sterling. It is clear that his intent was to eliminate the Mi'kmaq people from Nova Scotia, either by death or pressured exodus. The same man who represented "civilization" recognized that there were two peoples in Canada: the Europeans and the savages. He was also heard to say he would like to see the head of every Mi'kmaq at the end of a sharp wooden pike.

The same bogus treaty of 1713 allowed the British to scour the coasts of Africa in their modern ships – already equipped to transport a steady stream of human misery – and enslave, persecute, hold in bondage, and make huge profits from a people who were called animals, dumb creatures, because their skin wasn't white. With one stroke of a quill, for the first time in history, the Europeans had made the slave trade legal.

6

They sailed with the setting sun behind them, from the eleven-year-old British town now called Halifax. Its name as a French fishing village was already forgotten, as was its Mi'kmaq name, Chebucto. It was the year the white men called seventeen hundred and sixty.

Shouted orders and raising sails resounded across the big harbour, until that noise soon quieted and the schooner slipped out on the receding evening tide. Lying under the leech of the mainsail, Soulis Joseph was hiding more than resting. All the crew knew he was aboard, but the constant barrage of foul language that followed the shout of his name, accompanied with a few well-aimed kicks from some of the seamen, kept him out of sight whenever he could manage it.

During the sailing preparations while leaving the harbour, he had run from one side of the schooner to the other, trying to avoid the swinging booms and ropes and whipping sails as they were hoisted aloft. The hardy sailors always laughed at him when he scurried out of the way of all the activity necessary to get a fully-rigged vessel under way. Soulis Joseph knew nothing of the skills needed aboard ship. He had never been aboard one under sail before, though he had always wondered what it would be like to sail on one of the "wind catchers." His duties for his new master involved nothing more than tote-and-carry around the docks.

The men who had worked with him below the deck of the blackbird suspected he had seen the slave chained there. They had reported it to the brig's English captain. Later, when Soulis Joseph had been confronted by the master of the schooner, he said nothing; he just hung his head and bore the abuse spat at him, most of which he never understood anyway. The captain, who was now his new master, soon grew tired of it and in a fit of harsh coughing strolled a few steps along the port quarter. He laughed the incident off and told his crew what the British captain of the slaver had said about what the Mi'kmaq had seen.

"'Our legal and very profitable trade is frowned upon by many in these parts – though some love the cheap labour, mind ye – and they dress the blacks a little better, to make him more presentable, like. They're slaves jest the same, and came from the black continent in a ship jest like this one. Maybe even this one. Who knows? The black heathens all look alike.'"

Here the captain was overcome by another severe bout of coughing, during which bloody sputum exited his mouth. He retched several times and hawked a mouthful of phlegm toward the harbour water. It never made it. The intense coughing seemed to weaken him, but he continued on with his story anyway. He prided himself on mimicry and story-telling.

"'Still an' all, I'm not worried about what the bloody Injun saw below the decks of my vessel. He's not much more than a slave himself, an' jest as much a savage, fer sure. Can't even talk the King's English, he can't. Talks gibberish. Jest like the black niggers. They're all savages. The prime buck we have below decks will do well in my service in England. I'll fatten him up a bit on the last few days of the home passage. The oil from cod's livers puts a shine to their black hides, shows off their muscles, too. Ha! He'll fetch a pretty penny after I'm done with him. A sound investment all 'round.'"

He finished his parody of the English captain with a loud guffaw, which in turn ended with another fit of coughing. He walked forward to the yawning mouth of the forecastle, turned around, and climbed backward down into the shadowed companionway. Yelling for hot tea and still coughing, he disappeared below.

Soulis Joseph stared after the man who had seemed to enjoy his fear as he stood above him and cursed. He recognized some of the words, including his new name, Black Soolie.

He had been in the servitude of the French priests just a year before, and though he still remembered the sweaty, unwashed smell that always clung to the black, ground-length robes they wore like women, they never cursed on him and rarely beat him. Soulis Joseph had been with them for years. He wasn't sure how many. It was the French priests who called him Soulis Joseph. He had been told by his father that the salmon had come to the rivers fifteen times since he was born. He also knew that he had been given the proud name Amgwesewa'j by his father, meaning First One, because he was the first-born in his family.

He didn't like being called Black Soolie. The white men made it sound ugly.

The priests tried only once to pronounce his native name, which rolled off his mother's tongue like a warm summer breeze when she called him. They made it sound like *weasel*, laughing at their own uproarious interpretation of his name. His people considered the weasel as vermin, and though they trapped the small carnivore, they refused to eat its stringy meat. He never repeated his native name to them again.

It had taken him days to answer to the call of Soulis Joseph. Now the name Black Soolie was shouted at him as if he were a dog. He was a full-blooded Mi'kmaq Indian, his veins coursing with the spirit of a people who had walked this huge land for more than eight thousand years. While the European whites had been in constant squabble over puny tracts of land in Europe, the Mi'kmaq people had lorded over a land mass that could swallow the entire European nations as if they were nothing more than a sudden squall of summer rain.

The bays and coves of the ocean were no stranger to Soulis Joseph. He was already adept in the use of the canoe and even the French shallop. He was unprepared for the constant, sickening roll of the schooner, though. Kept in poor trim with insufficient ballast – heavy rocks, most of which he had wheeled aboard and loaded below in wooden barrels – the vessel lifted and just as easily poured down the slope of every grey Atlantic roller that came at her starboard side. He wasn't allowed below, nor did he want to go. The fresh sea air and the sight of the green land they were passing to port were the only things keeping him from vomiting.

Soulis Joseph huddled among the lashed gear. Tied to both sides of the heaving deck and laid fore and aft to the schooner were a dozen or more pine masts. He leaned back against the unpeeled fifty-foot trees, relishing the smell they carried of the land he was leaving. His new master had no way of knowing that the young man the crew now called Black Soolie longed for this trip across the ocean. The terrible nausea that rose from his stomach and which he tried to swallow back down, fearing further abuse from the rough sailors, could not prevent him from taking this journey.

* * *

Soulis Joseph's father was an independent and strong-willed man. Many of the Mi'kmaq rebelled against the domination of the French and their gradual stranglehold over the entire Indian nation. His father was one of them.

The day came when he had returned from his trapline almost empty-handed. One beaver pelt hung from his shoulder. He told his wife that he had walked beyond his usual hunting grounds to find fur. He had seen it coming, of course, but now it was here. Amgwesewa'j listened; his father was very serious. The Europeans had infiltrated across the entire face of the Mi'kmaq lands, their steel traps taking the animals in greater numbers than was good for the future.

"These people," his father said, "have no respect for our ways. They take the last beaver from the lodges. They slaughter the deer in great numbers and stuff the tender meat in their barrels, then cover it with salt and let the meat fester and stink till the true men would never eat it."

Amgwesewa'j had never heard his father talk for so long. He was a man who spoke softly, raising his voice only when he told the old stories around their crackling night fire. Amgwesewa'j knew there was more coming. It was as if his father was telling of a great hunt and was reaching the final, exciting end.

"We are leaving this land," he said, his breath expelling as if with relief now that he had said it. "We will go to our ancestral homeland. We will go to Ktaqamk."

When he had finished his statement he stretched to his full height and looked down, daring opposition from his wife. None came. Amgwesewa'j was swelling with pride and could barely contain his joy. The tales and stories about Ktaqamk were etched into his young mind. It was a place revered by his people. A huge island nation across the sea which gave them the sun. His people had been going there for generations. It was a place of great wealth. Countless deer ranged over its highlands and through the great forests which grew down to the very edge of the ocean. Fur-bearing animals always came to their traps. Great shoals of fish ringed its shoreline and the countless rivers there smelled of fish, they were so plentiful.

It was also a place of great danger. Other natives, not of their tribe, lived there, too. They painted their skin a deep red colour, not only for special ceremonies, but all the time. They were feared and respected. But Ktaqamk was a huge land, so large that few of their people had crossed the whole of it. They could live in this immense land with the Red People, as their people had always done.

Getting there presented a challenge and would have to be done with much searching of the skies for bad winds.

Amgwesewa'j knew from all of the stories that the crossing could only be attempted on fair pre-dawn mornings from the outside sea island his people called Kewtuminege.

Amgwesewa'j was so excited he didn't hear his father's new voice. What had he said? It could not be! Had he changed his mind about going already? When he spoke to him again it was as if a great weight had been placed on the boy's chest, so heavy he couldn't breathe.

"You cannot go with us, my son!"

He had never seen his father weep before. His mother's face was turned from him, her brown hands gripping the slanted poles of their teepee as if she were afraid of falling.

"The black-robed shamans who teach you about their gods and their many spirits have signed one of their papers. It says that you belong to them, and if I take you they will hunt us down and take you back, even if it means killing all of us. They say these marks on their paper are the law of their king. It must be honoured. Already they plan to scratch my name and my woman's on their scared papers. We will be forced to beg at their tables like the scrawny spotted dogs that always follow them, sniffing at their filthy, dragging robes.

"I will not be a slave to the white man, my son. My heart is heavy as stone to do this thing. We flee from our land and will reach Ktaqamk before we too are signed on their paper. If this happens there is no hope for us. From the sacred land of our ancestors we will join others already there and plan your escape to freedom."

His father finished his long talk, the last words barely audible, such was his pain.

Amgwesewa'j was crestfallen. He knew nothing about paper laws or distant kings and their cruel dominance over others. He knew only that his parents would go to the land of legends and he would have to stay with the hated men who not only talked funny, but wouldn't allow him to speak his own tongue. He fled to his mother's side, clinging and sobbing below her narrow waist until she couldn't hang on any more. She folded down to the packed earth floor in the centre of their teepee, taking her young son with her, and mother and child wept against the smoky, greasy poles.

* * *

His father and mother had hated the French military. They tolerated the French missionaries of Louisbourg, where they lived, and marvelled at the stories of their many spirits. Amgwesewa'j too feared the white man's spirit.

The priests told him stories they read from a large black book – one that he was not even allowed to look into – about their fierce God. A God who did nothing to save His own Son from a cruel death, because He knew he would come to life again. The Son of this all-powerful God could heal the sick of any ailment, bring sight to sightless eyes, and heal crippled limbs, yet had never saved any of the white men who had died of the terrible coughing sickness, despite the night-long pleadings from the Blackrobes. Amgwesewa'j understood very little of it. Still, he feared the God of the Europeans. One of the gods was a lesser one, but very evil. So evil that, if you chose not to follow the good God, the evil one would take you and keep you in an endless lake of fire where you would not burn, but where you would forever be tortured in unspeakable agony.

Amgwesewa'j never told anyone, but he feared the good God as much as the evil one, for how could he ever be sure he was pleasing the good God enough to prevent everlasting damnation from the other one? It was very confusing to him, but the endless tales of battles and great magic and countless mysteries the priests recounted from their sacred book reminded him so much of the spirits of his own people that he listened intently, sometimes wondering if the priests held tales of his own great Manitou hidden somewhere in their book.

It took nine terrible days of fighting in 1759, during which time as many as sixteen thousand British troops poured out of one hundred and fifty ships anchored in the harbour of Louisbourg. Though the French fought a valiant fight, it was no use. They were outnumbered and were defeated in short order. This time the British would make sure the fort would never again be used against them. They demolished the stalwart walls and burned the village to the ground while the French survivors fled for their lives.

At the start of the battle, Amgwesewa'j had hidden deep in the surrounding forest along with several more of his people who had escaped the besieged fort. They kept to the woods long after the sounds of fighting and the smoke from the destruction had faded, until at last the pangs of hunger drove the starving little band from hiding.

The British soldiers rounded them up and held them in hastily built compounds. The prisoners were thrown scraps of food, which they shared. Then they began a forced march to

Halifax along with several French captives, their trek in a few places aided by boats which took them across bays and still farther away from their land.

It was here in Halifax that Soulis Joseph was enslaved again and was now the property of the captain who called him Black Soolie. His name was always followed by an English curse. He had survived the leaving of his parents and the siege of Louisbourg. Three years later he was still working, and waiting for a chance to be free.

7

"We were always considered thieves and lazy good-fer-nothings, b'y," said Al. "You know, years ago, I mean. It's an awful thing, you know, especially when you've worked all of your life and have never taken a thing that wasn't your own. That feeling is gone now, thank God. Well, I hope 'tis gone. But it was there then and it took years for that attitude to be dispelled. I don't know why we were stuck with such a label. The Beothuk, now, they stole from the white man who invaded their shores. Maybe our people did, too, years ago. To tell the truth I wouldn't blame them. Imagine a crowd of foreigners coming here unannounced and, without any permission or even discussion, taking away their land and the animals and fish they depended on for their very survival. They were invaders, o' man. Wherever they went the white man took what wasn't his to take. And killed innocent people to achieve their ends. What do you think you would do under such a threat to your way of life?"

I agreed with Al, of course. It wasn't a hard sell for me. I had always felt the same way. In their passion for more lands, the white Europeans' domination was global. Native rights be damned.

We sat on one of the high green banks of the Shanadithit Brook north of Red Indian Lake. From my pack I pulled a bottle of warm water. Al gulped down the blue liquid named after – of all things – a gator. We were talking about the river below us, how it got its name, and the ancient people who followed the course of this river system. It was always our favourite topic and usually led to a long conversation. The Shanadithit Brook was wide and rocky. It was unusually high for midsummer. June had been a rainy month. Upstream and down it was much the same, strewn with rocks large and small, each one cresting the surface and tuning the river with a white bubbling melody. There was no defined channel that we could see, no clear way through the rocky route.

"I can't imagine gettin' a canoe along here, especially one made of birch bark," I commented to Al.

"Mmm. No, don't look good for a boat of any kind, now, does it?" he replied in a quiet, reflecting voice. "They did it, though. Walked them along, I'd say. Lugged them in places, then paddled and floated them in the good stretches. When they got to the lake, now, that must have been really hard. Digging them paddles in for miles and miles. Sure, 'tis almost always blowin', and the wind couldn't always be in their favour."

We were both silent then for a long while, which was unusual for my buddy.

I leaned back among the aromatic ferns and rested my back against a huge white birch tree. I wondered if the old Indians who had traced along these wooded banks had sat in the shade of the same trees. *You are too romantic*, I told myself. The life of a white birch tree seldom spans one hundred years, and even if a few of the Beothuk were still around here at that time, this tree wouldn't have been big enough to give much shade.

Maybe, though, I reasoned – refusing to let go of the possible connection – the seed from this tree had dropped from a similar one under which a tired native had rested, studying the rapids below and plotting a way through in his fragile bark.

We walked farther upriver from the woods road where the steel bridge had crossed over, and were now about a kilometre or so above the road.

"The Indians before us drank the water from this river and the cool streams that run into it, and here I am carrying store-bought water around with me wherever I go. Maybe we haven't advanced as far as we think we have, Al, b'y."

Al pulled the half-full blue plastic bottle away from his mouth with a sucking sound.

"One thing for sure, they didn't drink anything blue. Not only that, much as I loves the stuff, I hates havin' to suck from the bottle like a child." Al followed his statement with his usual chuckle.

It was late summer and over thirty degrees, not a good day for heavy walking.

"Too hot for Newfoundlanders," said Al. "Keeps the flies down, though. The buggers can't stand the heat or the cold. You know, when a feller sizes it up, for a tiny creature that needs ideal situations in order to survive, all of them bitin' insects make a good living. Yes, sir, the third plague of ancient

Egypt is still around us," he said, giving his arm a slap and killing one blackfly that didn't seem to mind the heat. Al reads the Bible a lot.

"To boldly go where no man has banged rocks before" is how I put it when he told me of his secret wish to prospect on the moon.

"Yes, sir, an' that I would. Right now, heart stents and all. Just imagine. The first man to look for minerals up there. No trees, no alder beds, no overburden of any kind. And best of all, no staked claims. A man could search around to his heart's content with not a living soul around to bother him." He was very serious, I could see. "I wonder how the ring of the hammer on the moon rocks would sound up there in that atmosphere?"

I marvelled at his dedication. He loved his trade. This tramping around, always in a searching mode, was his greatest pleasure. All of his family seemed to have the same love for this way of life. They had a special energy for "bush walking": no bog wet enough, no alder bed dense enough, no slippery, boulder-strewn brook with its constant halo of blackflies miserable enough to faze them from their seeking way of life. The Keats family plunged through the growth with packsacks bobbing and heavy rock hammers in hand and disappeared into their trade. They are as eclectic as they are steadfast in their quest. They don't all look alike, or in some cases resemble each other. But in their dedication and their love for the mysteries all around us they are single-minded, their abilities uncanny.

The Keatses are now a sixth-generation family of Mi'kmaq prospectors. Few family businesses can boast of six continuous generations of work in the same field. Maybe it's their love of the forest and the solitude it gives upon leaving the well-beaten path. Still, there are a great many of us who love the forests and being "away to the hills," and who are great hunters and exceptional woodsmen – you don't have to be Indian or even part-Indian for that. There is something more. In the case of Allan Keats, whom I spent the summer with, the difference wasn't easy to see, not even when I looked for it. It is more than travelling along endless forest access roads on quads, or walking among hills and precipitous ridges in all kinds of extremes, all the while searching for the unknown.

Al wouldn't appear to be noticing anything for as far as twenty kilometres along a very bad woods road. By this time my attention would be on other things, like animal signs or the great stands of timber we were passing, or, more likely than not, how to put all of this to paper. The pursuit of minerals was for the moment forgotten. Al hadn't stopped for the longest while. Then, all of a sudden, he would stop his machine unannounced – I learned early on not to travel too close to him – and hurry away, hammer in hand, and say, "That's the kind of rock we're looking for, my buddy."

I'd join him, always eager to find something new, and I'd be amazed. I wondered how he could stay so focused the length of a very difficult trail as to find a rusty rock, no bigger than my fist, beneath the scars of traffic on the old logging road. I asked Al how he could do it.

"I don't know, b'y. All I know is that it is something I've always been able to do. Maybe you're lookin' too hard," he added.

"Well, now, how can a feller look too hard for something he is constantly searching for?" I wondered aloud.

"I don't know if I can rightly explain it to you, but it's like a feller out moose huntin', especially a greedy feller. You know what I mean. One of them who got to have everything. You knows the type I'm talking about. He's hunting fer moose, all right, but he wants to get one so bad, and hates to have it said that he didn't get one, that all he sees are trees and bogs and endless cutovers. He is so engrossed in his search that he isn't hunting at all, just staring through the woods fer moose. He missed seeing the two ears that were twitching and listening to his heavy breathing and all the noise he was making from walking too fast.

"It's the same looking fer minerals. You have to pay attention to the little things. You know, like in life. The little things are the key to bigger things. I guess what I'm saying is, if you just be aware of what you seek, most things will come to you. You don't need to burn your eyes out staring fer stuff, moose or minerals."

He wanted to know if that made any sense to me. I told him that it did, and of course I meant it. It was like the old saying of not seeing the forest because of all the trees. From that day on I didn't search for minerals the same way. I just spent my time enjoying the country that was ours, and more often than not I found many wonders that had been there all along, even some minerals. I had just been looking too hard.

8

Flowing along by the south coast of Acadia, the schooner sailed east, favoured by a light southwest wind pushing at the starboard quarter. At regular intervals Soulis Joseph was startled by a flush of cold sea water that flew over the rail and raced across the swaying deck before it was sluiced out of the lee scuppers. He wrapped himself in a ragged black blanket. It had been a gift from the Blackrobes. That and a well-used knife, which he carried at his side, were his only possessions.

The hoarse cry of "Sail ho, jest off the starboard bow!" from the bearded helmsman brought the man the British seamen called Black Soolie out of his thoughts. He got to his feet and discovered his legs were as unsteady as a toddling child's. Holding onto the ropes that held the pine sticks, he peered in the direction the sailor was pointing.

Sure enough, looming high on the tops of the rollers with all but her topsails disappearing as she wallowed in the grey troughs, was a brig. The same one that he had been`aboard yesterday. The same one that held the black prisoner, of that

Soulis Joseph was certain. His keen eyes identified the slaver at first sight. The others, too, all on deck now and staring to starboard at the brig their schooner was overtaking, knew the scorned ship.

"'Tis the bloody blackbird, by gawd," came from one of the seamen.

"Slowed, she is, with all of them lovely full rum casks aboard of 'er," said another.

"There be no rum comin' from the 'old of that one that I'd quaff across my lips, sweet or no," stated the wheelman.

"We'll soon be gettin' 'er smell now, we being downwind an' all. She carries a stench worse than smatchy fish."

Soulis Joseph listened to the banter coming from the sailors and watched the nearing vessel, wondering if the *antaqvej* was still aboard. He shrugged his dirty blanket a bit tighter around his shoulders. He didn't remember seeing a blanket anywhere near the chained slave. It was cold in the bottom of that ship, a damp, penetrating chill that rose from her foul-smelling bilge.

The smell from the slaver was now full across the schooner as the two vessels passed.

"By gawd, if they flushed that one in the Indies 'tis a poor job them warm waters 'ave done of it. They should flush the bugger up here in our icy waters," remarked one of the sailors.

* * *

The brig had come from the British Isles like hundreds more before her. She had slipped out into the stream at

Liverpool and went with the low tide and a favourable wind away from the crowded land and had begun the first leg of her Atlantic triangle: the outward passage.

She was loaded with bright, coloured blankets, barrels of beads and cheap baubles, looking glasses and ribbons, and bolts of cloth that would hold no warmth, all of which had been purchased with sweaty shillings through the back doors of a Manchester warehouse, long after business hours. Gunpowder, British muskets, iron nails, and knives forged from the dregs of the steel mills were all valuable assets for the purchase of man's vilest commodity, the flesh of his fellow man.

South she had sailed, following the warming seas of the grey Atlantic, until the great bulge of Africa brought to her port beam the warm perfume of a mysterious land. The sea had all of a sudden become an unbroken blue, luring the merchant ship onward. Crossing the invisible line of Cancer, her forefoot plunged on into the tropics, past Cape Blanc. After passing Cape Verde and heading a bit east from her southerly passage, her crew caught their first glimpse of the Dark Continent full on their port bow. They had reached the Slave Coast of Africa.

Here in the gulf of Benin, on Africa's west coast, her foul deed began. Hundreds of black-skinned men, women, and children alike were waiting to be purchased, many of whom would be trapped and held in great stinking barracoons by their own kind, and sold into a lifetime of degradation and unspeakable misery. World leaders not only knew of their plight but condoned it.

It would take almost a hundred years before the wretched business of buying and selling humans would be made illegal by the same world powers who had made it a thriving industry. By this time several millions of African people would be transported across the South Atlantic to be transplanted into the indentured workforce of the fledgling American colonies.

The middle passage of the slaver's trek started with a well-loaded human cargo, all stowed out of sight below decks. Borne along by the southeast trades as it neared the equator on the fringe of the South Atlantic, the blackbird gathered itself for the northern shores of South America, polluting the air with its smell every time it rose above each favourable sea. When the slaves were forced on deck at rare intervals, even the young appeared old, as their cramped muscles wouldn't allow them to walk upright on a rolling deck. They were doused with cold sea water in a useless effort to cleanse them. Along the coast of Trinidad to port and the isle of Barbados on the starboard side the vessel sailed into the blue waters of the Caribbean to find a ready market for her chattel.

After the crew off-loaded the slaves, they purged the vessel of human waste. The captains and owners of these ships had come up with an ingenious method of accomplishing this. They stripped the slaver of everything movable, including all the belongings of the captain and the crew. Cordage and spare sails were off-loaded, as well as every item that could be damaged by the salt water.

Then the crew brought the ship under full mainsail to a deep, cliff-lined cove with a sloping beach. Here they drove the blackbird full aground upon the strand at spring tide. When the tide lessened and the vessel listed down onto its side, they lashed it to the rocks, so that it looked like a huge beached whale. With poles bared, it now awaited its cleaning.

The next tide flushed into the very bowels of the ship, as like a living thing its timbers creaked and strained to rise. This process was repeated on the following tides and on the opposite side of the slaver. The salt sea streamed into every crevice, getting rid of most of the filth, but it could never rid the blackbird of its smell. It had seeped into the wooden sides and soaked the hardened oak beams, so that when the ship was restored to sailing form again and looked the part of the noble sailing ship, its foul smell would betray it wherever it went.

The slaver was now ready to trace the penultimate line of its infamous triangle, to collect rum and sugar, tobacco and logs and lumber, all from the sale of exiled Africans. Then it headed home along the last and most profitable stretch of the voyage – the homeward passage – to begin the cruel process again and again.

By the time the slave industry ended, it had already changed world history. Never before in the annals of mankind had such a forced migration of an innocent people taken place. The numbers of captured and enslaved African people were greater than the combined tally of all the ancient conquerors. Even the famed "infidels" who had for years raided for slaves out of the Persian Gulf, or the dreaded Mongols and Vandals who scoured the great plains of Asia, had not forced their will upon as many defenceless people as had the Christian people who lived along the shores of the North Atlantic.

* * *

The British sailor's last remark went unanswered, as one by one the seamen on board the schooner turned their heads away from the slave ship and toward the port side of their own vessel. It was a scorn worse than shouted insults, for sailors the world over always greeted passing ships with shouted hellos. That the blackbird was held in such contempt by fellow seamen was flung across the sea between them as loud as if it were bellowed from cupped hands. They passed in silence. Even the rigging seemed to have smoothed out its creaking.

Once, one of the sailors looked across the span of water between the two vessels, but a piercing glare from his captain spun him around as if he had been slapped. Now the only sounds were the dull creaks of the straining rigging and the luffing sails.

The captain was overcome again with a hacking cough. He quickly muffled his entire head with the hem of his black coat, daring to let no sound escape while his bent form trembled and staggered in an effort to restrain the pressure that wanted out of his lungs.

Watching the captain trying to silence his misery, Soulis Joseph remembered all too well the coughing disease he was sure the man suffered from. He had noticed it long before, when he had first seen his new master. The captain had the same disease many of the Blackrobes had carried. The White Plague, they had called it; their pale skins turned even whiter and more deathlike as the dreaded sickness worsened.

He had been forced to tend to many of the dying priests. Most of them were too weak to pray to their God. He knew the sickness could enter his own body as well as the white men's. It had killed many of his people while he watched in helpless fascination. But he had never been sick, though he had lain on the same bloodstained blankets as many who had died. He had no pity for the man who had enslaved him.

Only Soulis Joseph stared at the slaver as she slipped astern. He knew nothing about the ways of sailors, and less about the flourishing trade in human flesh that destroyed the lives of another native people a world away from his own. Ignored by the others, he watched the rolling brig until it was gone. He knew now that the *antaqvej* was still aboard the stinking ship. He didn't know how he knew, but he was sure the man was still on board, and still chained like a village dog. Soulis Joseph could feel his pain.

9

Black Soolie escaped from his white master, just two months after his arrival in the bay the white men called St. George's. He simply disappeared into the forest. He went searching for his parents, after learning from other Mi'kmaq in the settlement that people similar to the description of his parents had indeed been there. Someone also told him that the man he believed to be his father spent most of his time inland on a long trapline. His woman always went with him, the person said. Soulis Joseph hoped to find them somewhere in the hidden valleys.

After surviving his first winter in this harsh land, living by his wits and in constant fear of pursuers, Soulis Joseph made it to the deep interior of the island by springtime. He had not found his parents, nor any other living soul.

Once or twice he saw a few indistinct traces where trappers had been. The sharp, slanted end of a small tree stump near a beaver lodge, the cut clean, unlike the gnawed teeth marks of the beaver, had to have been cut by an axe. The long-dead ashes of a campfire close to a brook also spoke of human passage, but he couldn't be sure if they had been white or Indian.

He skirted these few places with great care, hoping to catch first sight of others from a distance, until he knew for sure if they were his own kind. To most of the native Indians, strangers were assumed to be enemies and were always approached with caution. Soulis Joseph didn't know how many enemies he would find in this land.

He had kept to one of the many river courses he followed, surviving on hares and salmon and the frozen squashberries which lined most of the riverbanks. On the night he escaped from his captors he had taken a rusty knife from the cuddy of his master's dirty punt. One day he had also discovered a worn flint and steel among the scattered stores in one of his master's sheds, and the filthy sleeping blanket that was his only night cover, he found in the corner near the drafty door where he slept. He had fled without stopping, for two gruelling days, until tired, aching limbs and an empty belly denied him farther travel.

The knife he kept sharp by scraping it across wet rocks. He tied the blanket around his neck and draped it over his back; the ends of it became ragged as it dragged over the bushes just below his knees. He used the roots of young black spruce trees as snares. Hares were plentiful and easy to catch. He seldom used the knife to skin the small animals.

Lifting the rabbit by the hind legs up over his head, he would fling it over his back. Then, with a violent downward motion, he would bring the hare forward and just as violently yank it upward. The sudden stop would dislodge the animal's bloody viscera and spill it at his feet in one fluid motion. He had learned this trick of field dressing the tasty hares from his father. This method not only sped up the gutting of the animals but reduced their weight when dozens of them had to be carried on a hunter's back.

He carried a sharpened, fire-hardened stick in his hand at all times. Twice he found spawning salmon. They were smaller than he was used to seeing in his own land. Standing knee deep in the icy water, he waited without moving until one of the fish swam near. It took several tries, but he was always successful in spearing several of them. They were dark-coloured and not as tasty as the ones that came from the mouths of the rivers where the shiny fish left the salt sea. Their hungry bodies were gaunt, devoid of the fat that gave them their rich taste.

He stayed at such places for days, gorging himself on the fish that he roasted over fires he lit with the flint and steel. Dozens of them he opened up and placed over poles angled over a slow, smoking fire all night long. He carried the smoke-cured fish with him when he moved on.

Soulis Joseph could see now why his people viewed this big island – this Ktaqamk – with such respect. He couldn't believe the size of it. It was endless. It was as if he were all alone in this vast wilderness, with only faded signs along the streams and rivers of a few hunters who had passed the same way.

His appearance by every pond and river pool would flush out ducks and other waterfowl. Great schools of fat trout and salmon showed in many of the waterways. By one of these riverbanks where he had spent days curing the red-fleshed salmon over a smoky fire, he saw two black bears that had also been feeding on the spawning fish. On another day by another river, he was close enough to see the string of slobber from a bear's mouth as it gulped down the the belly of a large fish. But they ran when they saw him, their black hides shining and undulating. He had no fear of them.

Along the edge of the same watercourses were beaver lodges, with their attached pantries of cut alder and young aspen, whose sogged, yellowing leaves floated away from the rodent's tangle of winter food. Muskrat hubbles, their round, domed roofs bunching up out of grassy swamps, yielded several of their inhabitants to his hunting skills. He saw red foxes and once glimpsed a black one.

Lynx tracks followed the leads of hares that snaked through the forest glades, though he didn't see any of the wary animals. Twice he saw small herds of caribou. He longed for their rich meat, but though he waited, with spear in hand, for hours in one of their narrow leads that led to a long, high bog, no more of them came. Following one of the herds were the tracks of three wolves, their steady lope unmistakable. On that same night, as he lay beside his fire and waited for another muskrat to roast, he heard the distant wolves howl a successful hunt.

Swift, dark brown animals disappeared among the branches of huge trees when he disturbed them. Their fur was the richest of all. The small, shy animal was what the white men called the marten cat. Otters peered up over slippery rocks at the Indian's approach, diving back down and reappearing for quick glances as he walked by. Weasels dressed in their fine winter white pounced away, their black-tipped tails swaying.

Soulis Joseph would never leave this land again. His ancestors had not exaggerated. This was a sacred place. He could feel his bloodline pulsating through the pathless forest. This inexplicable feeling of belonging was like none he had ever felt before.

The young Mi'kmaq sat by a clear stream one day and drank his fill of the cool water. He had eaten well for days now. All of a sudden, as if he had just been welcomed at a long-sought-after door, he knew what this feeling was that he had been experiencing for days. The realization was so strong he thought he heard ancient voices seep through the trees all around him. He stared through the woods, almost seeing the spirits he felt. At the same time he shivered, though he was unafraid of their presence. His young heart pumped with a rare joy. For the first time in his young life, he felt a sure sense of belonging. He was home.

10

At the beginning of one long trip in particular, I was the cause of a great deal of laughter from Al and would bear the brunt of his retelling of the story many times.

We had left the cabin we were staying in very early one fine Sunday morning. It was already close to twenty degrees, as the modern thermometer hanging from the veranda outside our cabin door showed. It was a good morning to be alive and healthy and heading out on another adventure.

We lashed our gear to our faithful quads – we almost always called them bikes – and set out for another long day of searching for minerals. This time we were after an anomaly showing on Al's geology map, close to the centre of the claim we were assigned to investigate. We had studied all the details the night before and, after much discussion as to the best way to get there, Al plotted out our proposed route on a topography map.

The first part of the trip was over a thirty-five-kilometre, grown-in and rocky old logging road. We would have to work inward from the shoulder of this road and cut a way through more than several hundred feet of dense alders with axe and machete to allow us access onto a series of boglands. According to the map and our calculations, the end of one of these marshy leads would leave us with another three kilometres or so to walk to the top of the wooded ridge and the anomaly. It wouldn't be an easy traverse, but it had to be done.

Al was travelling ahead of me at a steady speed, picking the best route. The sound of his engine was like the drone of feeding bees, rising and falling as Al slowed and sped up again. His loaded bike always looked the same. He tied his prospecting gear in the exact same place every day. Lashed with several bungee cords to the back of his bike was his red packsack, a short shovel, a sharp axe, and a sturdy black rectangular box which contained the scintillometer, to check for uranium if Al saw the need. Dangling by its metal handle was a white, brown-stained plastic bucket which we used to carry gravel to the nearest water to pan for gold. I always figured it would chafe loose from its moorings and fall off, but it never did. On some of our forays through dense bush – alders, for the most part – the only part of his bike I would see was that swinging white bucket.

On this day he stopped without warning, as he always did, and said in a loud voice, "Oh, luh! God love 'un. Luh."

A male spruce grouse stepped out, neck outstretched, proud as any well-adorned peacock, and walked along by the port side of Al's machine. We kept pace with the bird for several minutes until we both passed it. We sped up to get away from the friendly grouse, fearing it would get run over. Al kept looking behind, standing on the running board of his machine, and twisted my way.

"He's still comin', b'y," he called out in his laughing voice.

I looked behind, and sure enough, the friendly grouse, its neck outstretched like a rutting rooster, was still racing after us. I went faster.

Al's cry of "Look out, he's dive-bombing ya!" was almost lost in the noise of a crashing blow against the back of my skull, followed by a fluttering of brown feathers and silky down falling all around me. For a moment I was puzzled as to what had happened, so sudden and violent was the attack. Between peals of laughter – I thought he would lose his breath, he was laughing so hard – my partner soon enlightened me.

"The grouse attacked you, b'y. In all me days in the woods, I've never seen the like of it." He paused for breath, then laughed some more. "I 'lows he's a she and took a liking to your head. They like shiny things, you know."

And with that, Al left his bike and keeled over in the trail, beside himself with mirth. All this time I said nothing, half expecting the killer grouse to appear again, more savage than ever. The bird was nowhere to be seen, but Al wasn't finished with me yet.

"I finally knows why" – he couldn't continue for several moments because of his boisterous laughing – "why the government made it a law to wear helmets on bikes. 'Tis to prevent injury during grouse attacks!"

He roared all the more at his own joke and doubled over again and again as tears rushed down his brown cheeks. My laughter soon joined his, and we laughed uncontrollably, there at the scene of my very first grouse attack. I told Al that the bird looked like a he to me and that if I could catch the bugger I'd have fried rooster for supper. I went on to say that I loved breast of grouse and would relish the taste of my antagonist.

"Ah, God love 'un, sure. 'Tis only revenge you're after. Still, the breast of that one will be well-tenderized after slamming into your big head." His body shook as he tried to draw breath between bouts of laughter. I would never live this one down.

We finished our trek that day without finding the anomaly, and when we returned to the scene of the attack on our way back to camp, he of course stopped.

"Be careful, now, this is a dangerous area. Put your helmet on. Ha ha! Want me to strap it tight fer ya? Them grouse can be killers, ya know, especially if they gangs up on ya. Ha ha ha! Keep as close to me as you can, now."

And he laughed and laughed again.

We did see the grouse again that day, and though it ran along by the side of our bikes for several hundred feet as before, it didn't attack. And when we sped up to pass the bird – Al had been driving real slow, hoping the bird would repeat its morning efforts – I warned Al that "Grouse don't always strike twice on the same head, you know." He roared with his deep laugh, dwarfing the sound of his quad's changing gears, while I raced after him with head bent and shoulders hunched, as though I were about to be attacked from behind by a charging bull moose.

One of the amazing things I discovered about Al very early on was that he hadn't spent a lifetime searching for rare metals for any great financial reasons. It was a great way to make a living. Even when we found flakes of gold at the bottom of his pan in a small stream, it still wasn't about money for him. It was the thrill of discovery. He could hold the secret of a rare find forever.

Having his name attached to a mine discovery, like the big one at Duck Pond near Millertown, was unimportant to him. When I'd mention his major role in that find he would always stop me by saying, "Naw, b'y, there was a crowd of us who worked at that discovery."

Al would be most proud if in his work he could only discover a rarity – and he is convinced that it is out there somewhere – that would be a breakthrough in the medical field, particularly in the aid of sick children. He was always convinced that plants or minerals held the secrets that would cure all major diseases. Man had only to find them. "Maybe you and me will find it today," he would say when we talked about the possibility.

The feeling of pitting his wits against nature and winning in some way – and he has done so several times – was a great pleasure to him. In 1975, Allan Keats and his brother Fred, as well as their father, Ted Keats, were working for Noranda, one of the largest mining and exploration companies in Canada. They were investigating rock formations and sampling their finds in the Millertown area. What they were looking at was a Buchans type of ore body, massive sulfides with twenty to thirty per cent of copper, lead, and zinc showings.

Their discovery would wait thirty years or so before it came into its own. Ore prices, as well as the demand for metals, would not show the proper returns on the ledgers of mining companies until the turn of another century. By that time the huge Noranda mining company had long gone from the island of Newfoundland.

With the opening of the Duck Pond production mine, the tiny community of Millertown came to life again. Mighty trucks roared to the West Coast of the island, bearing the crushed minerals away from their forest home. In Stephenville, the mineral was loaded aboard huge ore ships in the dredged port and transported to mainland mills, where it would be refined and sent to world markets.

Millertown's real beginning had come with another industry. Lewis Miller had a thriving sawmill operation in the coastal town of Botwood as early as 1890. He expanded his business to the southeastern shore of Red Indian Lake at the start of the twentieth century and began Newfoundland's first significant logging and sawmill operation. The cluster of logging camps, sheds, and horse barns came to be known as Millertown. The town has kept the name of its founder even today.

I asked Al what reward, if any, he had received for the discovery at Duck Pond. He prefaced his comment by saying, "It wasn't about a reward, you know," but by this time I already knew that.

"We had a job to do, that's all. And we did it in the best manner we knew. Like we always did, and still do. We were looking for minerals for that company, and we found them. Wonderful. That's what we were there for. We were paid for our work, and if they wanted to give us something extra, well an' good, that was all right, too."

I knew Al had found the first significant boulder containing valuable sulfides. When I asked him about it, he was quick to remind me that it was a joint effort.

"A lot of us involved in that discovery, b'y. Not only me. Fred and Father, as well as the mining engineers. Everyone plays a part in a find."

"The something extra you talked about," I asked him, "what was that?"

"Oh, I think Fred and me got four or five thousand each for it." He chuckled.

I reminded him of the gratuity old Matty Mitchell, the famous Mi'kmaq prospector, had received from the Anglo-Newfoundland Development Company for his huge discovery of the Buchans ore body. For his discovery of one of the largest bodies of sulfides anywhere on earth, he had been given one barrel of flour. In 1905, when Matty made that tremendous find, a barrel of flour wasn't worth much more than a couple of dollars.

"You can buy a lot of flour for a few thousand dollars, even at today's prices," Al said. "Still, I know what you mean. By comparison it wasn't much more than Matty Mitchell got. But like I said, it wasn't about a reward. It was about doing our job and doing it right. Not to mention the wonderful feeling that comes from being the first to find one of Mother Nature's mysteries."

11

As he travelled he searched for signs of his own people. He knew they were here somewhere. Bits of information he had understood from the white men at the coastal premises assured him there were many Mi'kmaq trappers inland. Most of them brought their catches back to the village at St. George's Bay. His master depended on them for his winter's supply of furs.

Once he stared at a greenish, rusty stain that oozed out of a massive outcrop of rock. Some long-ago erosion had exposed it from the forested ridge he was crossing over. Standing below the cliff that disappeared inside the treeline above, he spotted a vein of mineral different from the rest of the rock formation. Embedded within it were shards of a glittering yellow rock. He poked and prodded at it for a while, enjoying the lustre of the tiny yellow rocks, but he could see no immediate use for it and walked on.

One day he found chert, his eyes always falling on the things that were not of the ordinary. Now the worn flint he had stolen wouldn't be a problem. The chert he had found, unlike the grey flint, was the colour of rust. His peripheral vision had brought him to it. It was nothing more than a tiny red piece of stone which had fallen free from the escarpment he was passing, but it had drawn his attention as if it were a mountain of red rock. Searching, he found many more pieces.

Now he had all the fire starter he would ever need. He knew of other rust-brown rocks he could use to strike against the chert to create the spark needed to build a fire, but he would hang on to the steel piece he had taken from his old master as long as he could. Not only could he make fire with the chert, but he could also fashion a crude axe from the material, enabling him to cut deeper into wood than his knife allowed. Feeling proud of his find, he placed several select pieces of the chert into his makeshift birchbark bag and moved on.

Soulis Joseph made his way along a riverbank to a place where the swift flow ran away from a large pond, or maybe a larger lake. He picked up his step, eager to see where the river he had been following came from, when he stopped, as if he had been ordered to do so. Voices! Loud and clear on the evening air.

To his right and below the faint trace he followed, the river ran dark and swift, almost silent. The voices seemed to be coming closer. He was filled with a sudden fear that for the moment overwhelmed him and he fled into the forest, running until he was breathless. Long shadows had bedded down, and stars shimmered above the treeline before he decided to sneak back the way he had come.

Maybe the voices he had heard where not those of the white men he was fleeing from. Maybe they were his own kind, or even his own parents. What had he been thinking, to flee like a child from what may be the one chance of finding his parents? He had to go back and see for sure who owned the voices.

Under the shield of night he crept, until he reached the edge of the riverbank where he had first heard the voices. Then he moved a little farther on, toward the place where the water left the pond.

He smelled the smoke long before he saw the campfire. He stopped. The smell of a burning campfire in the still, cold night stirred his emotions. Fire was the very symbol of comfort which had drawn Man in from the cold since he had stood erect; it burned as it warmed his naked skin and kept away the dark.

Soulis Joseph heard the voices again. Soft sounds. He wasn't close enough to understand the language. They were in the same place as before, of that he was certain. They hadn't been coming closer, as he had feared. The soft evening breeze had tricked him.

The wind, though light, was bearing away from the campfire and toward him. He smelled roasting meat. Caribou meat. The rich aroma wafted among the trees in search of his nostrils. His mouth watered.

He crawled close enough to hear the fat from the meat bursting as it dripped into the fire. Creeping closer still, he knew the sounds from the river tumbling over the rocks would disguise some of the sounds he might make. But he also knew that a broken twig would pierce the darkness with an explosion of sound, and alert the people around the campfire as surely as would a shouted greeting.

Where the smooth flow of water left the pond and hurried through the opening in the forest, it dipped away from the black pond before it narrowed and formed a V that ran around the nearest rocks. The river swirled into a large black pool before it disappeared around a sharp bend and out of sight. Here, where the rushing water made the most noise, Soulis Joseph waited and watched the campfire. He was close enough.

There were three people around the modest fire, and although their backs were to him he thought one of them was a woman. They weren't white, but Indian. But would they be friendly to him, or would they betray him? Soulis Joseph wasn't sure what to do. He decided to wait and watch for a while.

The cooking fire gave little light, but he could see two of them were adults. The third was much smaller than the others. A child. A girl, he figured. A cone-shaped wigwam built of dirty white birch bark stood nearby. A leather strip held a dried caribou skin back from the dwelling's entryway to reveal a dark interior. He worried that there might be more Indians inside the shelter.

Soulis Joseph decided the tall man was a *gtantewiau* – a hunter. He had been right about the smell of the roasting meat, too. A full caribou rib cage, supported by blackened forked sticks on either end, hung over the low fire.

Staring at the scene before him, Soulis Joseph avoided looking into the fire itself, knowing it would rob him of his vision. He watched the man and the woman most of all. They were each occupied with a task spread before them. The woman was sewing a heavy skin garment, while the man was rolling what appeared to be a thin root across his knee over and over again, as if he wanted to soften it for further use. After rolling it several times, he pulled the white tree sinew taut between his outspread hands, only to lay it down and begin again. The young girl stood when the man spoke to her and began gathering short pieces of driftwood in the shadows beyond the campfire.

Soulis Joseph tried to recall his mother's mannerisms. Her way of sitting, how she walked, the sound of her voice. Could this be her? Could the small girl be a sister he never knew? The man who was making the twine was focused, bent over his work. His father had always been that way, giving a needed task his entire attention. Could it be them? he wondered. In this vast land of Ktaqamk that he had been searching through, had he found what he needed most? He shivered with the anticipation of meeting with his mother and father again.

The woman rose from her crouched position and approached the campfire to turn the searing meat. She couldn't reach both ends of the skewer at once. The man, who had been watching her, got up, grasped the other end of the stick, and helped the woman turn the meat until she was satisfied.

The hunter returned to his place a few feet from the fire, while the woman stayed and watched the simmering meat. They continued to talk in low voices. It was a pleasing, comforting sound. So far Soulis Joseph had heard only murmurs from the group, none of which he could understand. Snatches of conversation that he did hear sounded like his own language.

The cooking fire didn't give enough light to see them properly. The low, flickering light of a campfire will often change shapes within it. Ribands of light seep through the forest as if they would dislodge the dark. Moving objects, such as flitting moths, which dare near the light are distorted and appear larger than normal. Sometimes the faces of the Indians he watched appeared grotesque, while their shadows danced and high-stepped behind them.

The campsite had been well-chosen. A circular cove adjoined one side of the river mouth. A sandy beach led away from the quiet cove and tilted upward toward the treeline. A deep-sided canoe drawn partway up out of the quiet water pointed its bow high. Its stern disappeared in the shadow beyond the light. The canoe, like the teepee, was made of the bark of the white birch.

The ground levelled where the sand ended and the trees began. Tall grasses and sedges grew among the fringes of the thickening trees. Here the lone teepee stood sheltered, almost invisible from the river or the pond.

A word from the woman bending over the fire announced the meat was ready. The man discarded his work and hurried to her side. The child, who had been bending in and out of the ringed shadow created by the cooking fire, appeared with her arms full of white, seasoned driftwood.

High above in the dark sky a snipe searched, its wings emitting their wobbling hunting sound as it plunged after fleeing winged insects. From somewhere behind Soulis Joseph an owl hooted. Farther away, on the other side of the dark river, another one called. All this he registered as he watched and waited.

With the aid of her husband, the woman removed the meat from the fire and placed it, still simmering, on the ground. The man spoke another word and turned his head toward the child, who hastened to the glowing fire. She placed several pieces of the driftwood upon the glowing embers, which burst into bright flames in an instant. They illuminated the campsite and filled Soulis Joseph with terror.

The sudden flash of white shone brighter than a winter's full moon. For several yards around all was a stark glare. The light flashed full on the girl-child who had collected the firewood, revealing her face. She was Beothuk!

12

Soulis Joseph dropped and spread himself prone upon the ground, squirming deeper into the bushes. He wished he could disappear. His sudden, panicked movement had made some noise, he realized. He dared a peek up over the edge of his concealment.

The hunter was looking in his direction. Soulis Joseph tensed. Cold spasms ran down the length of his lean body. He was shivering. His teeth chattered.

Just then the owl on his side of the river jumped away from its perch in a loud flutter of wings. The Red Indian followed the sound of the hunting bird until it faded away, and he turned toward his waiting meal.

Soulis Joseph felt faint with relief. He had never known such fear as that which was now coursing through his veins. The stories of the red-skinned men who dominated the interior of Ktaqamk were legend. Their skin bled the rich colour of their fame. It was said they chopped off the heads of their victims with blunt stone axes. They even dug up the bones of their dead long after their burial, using only their hands, and painted the skeletons. It was said the stain from the skin of the living would stay with the bones of the dead even into the afterlife. The Beothuk would always know their own in the world of the Great Manitou.

They hunted the huge white north bear with their wooden spears on the drifting spring sea ice. Unlike the Mi'kmaq and the rest of the mainland native peoples across the water, they still fought against any form of association with the white-skinned Europeans. They were wary and elusive. Soulis Joseph couldn't believe he had found them. He had to remind himself that he hadn't been looking for them.

With only his head raised through the bushes, he stared at the family of Red Indians until his neck grew stiff. His body ached from the cold in the damp ground as it seeped through his thin clothing. Soulis Joseph ignored his discomfort.

His fears were calmed after a while and, as he watched the Beothuk, his confidence grew. He was downwind from the encampment and well away from the light given off by the campfire. Anyone looking out from the encampment would be looking into complete darkness and would see nothing at first glance.

Still, the hunter had heard the slightest of movements. This man would not be fooled a second time. Soulis Joseph decided he would wait until they entered the mamateek to bed down for the night before creeping away into the darkness.

He was filled with awe by the red-skinned people. He couldn't believe he was watching them! The hunter's face was a deep red and appeared to be darker than the others, as if the colour were weathered. Even his long-fingered hands were the same shade of red. The young girl's skin was a brighter red that seemed to shine when the firelight found the henna. The woman's colour, though also red, was a softer hue than the man's and the child's. It highlighted her full round face. Soulis Joseph thought she was beautiful. Maybe they all oozed their red blood in different ways, he thought.

All three were dressed in animal-skin clothing, the sides of which were slit up the legs to make for easier walking. Underneath these robes of skin they wore leather leggings. The man wore a furred hat which came below his ears and stopped just above his eyes, as if it were too big for him. The other two wore no head covering. Sticking out from under his hat and fastened to the mane of black hair just above his proud forehead, the hunter also wore three black feathers. They bounced and dangled down over the side of his head as he moved about. Soulis Joseph thought they looked like the feathers of a raven's wing.

The warning that the elders had given to the hunters who wanted to venture across the great *Gta'n* – Ocean – to this land frightened the young Mi'kmaq. He had been travelling in an easterly direction almost all winter long, sometimes a bit north. His mind flashed back to the stories of his people again, only now accepting how valuable they were.

"When walking in the land of Ktaqamk," they had said, "always remember, Red Man that side, Mi'kmaq this side."

The storyteller had pointed east for the Beothuk and west for the Mi'kmaq. Soulis Joseph figured he was too far to the north and east and had crossed the invisible line between the two nations. He stared all around him, as if expecting great hordes of axe-brandishing Red Men to appear out of the dark forest behind him, and he was afraid.

The family of Beothuk settled into their evening meal. The woman cut a section of the meat free from the roast and gave it to the child. The man cut off a large piece for himself and sat back on his haunches. The woman hacked off a piece for herself and sat beside the young one.

They talked as they enjoyed the meal. Soulis Joseph drooled with hunger. He could almost taste the caribou meat. For one fleeting moment he felt like revealing himself – anything for just one of those succulent ribs. However, he soon overcame his desire and resolved to wait. As he watched the Red Indians prepare to lay down for the night, he decided they were no different than his own people.

The man walked away from the light and urinated, studying the night as he did, his gaze all-encompassing, the whites of his eyes shining. The young girl left the ring of firelight before she too crouched, made water, and hurried back to the camp.

The woman got up to do the same after the man returned. She walked out of the light and came close to where Soulis Joseph was hidden. His anxiety rose again but he dared not move as the woman bent beneath the bushes and yielded to her need. She was so close to his hiding place he could hear the soft hissing sound her water made as it fell upon the ground. She stood to let her skin dress fall around her legs again and returned to the fire without looking back.

The woman was the first to leave the fire circle. She took the girl-child by the hand and walked toward the mamateek, then stooped and followed her into the cavernous night of their shelter. A few minutes later she emerged and joined her mate at the fire. They talked for some time, their tongue soft and caring as they spoke. Soulis Joseph thought he heard the hunter say "Teehonee" a couple of times and he wondered what the word meant. Maybe it was the woman's name.

Soon the woman got up and strode back to the mamateek. The Beothuk man scraped the fire coals closer together, placing the unburned ends of wood in the centre with care before adding several larger and much heavier pieces to the smouldering campfire. He looked around him, then went to the beached canoe to complete his inspection before retiring to the dwelling to join his family.

With one long leg bent inside the structure he stopped, looked around again, and sniffed the air. He waited a moment, then sniffed again, his head raised. Another moment passed before he loosed the flap of hide and crouched like a wolf and entered his den, letting the thick caribou skin fall into place behind him.

Soulis Joseph stole away from the camp of the Beothuk, but not before he heard the sound of snoring coming from the mamateek. By sun-up he was far from the site, having walked through the night. When the day came to the land, he turned and walked south, keeping the rising heat to his left.

For days he wandered the hills and valleys, fearing the White Men to the west and the Red Man to the east.

* * *

The next time he encountered Indians, they found him by a trail they were following, as he dozed in the heat of the day. His alarm was soon dispelled by the people of his own race, who spoke his tongue and who were overjoyed to meet him. He was welcomed to go with them and stay with them forever, if he wanted, south to a place they called "The Conne."

After he told his story twice to the trappers who had found him, they shared some news that was of great importance to Soulis Joseph. They did indeed know his parents. The trappers had lived with them at the place they would now take him. It was a remarkable place by a deep-sea bay where the whites seldom came and where a Mi'kmaq village was growing.

It was a place of their own. They would never leave. However, his parents, he was told, were no longer there. They had left, both of them, with their winter furs. His father was a great trapper, he was told, though he already knew, and had walked to the bay of St. George, where they always got a fair price for their prime fur. To make travel easier, his parents had left before the strength had gone out of the spring ice. They intended to sell their skins there and gain passage back to Nova Scotia to find their son, Amgwesewa'j.

His mother had grieved for him always, until his father agreed they would go back and bring him here to this sacred land. Soulis Joseph was saddened by this news. To have come so far and to have been so close to succeeding in his quest to find his parents, only to be dealt such a disappointment, was a terrible blow to the young Mi'kmaq.

Would his parents find out that he had followed them here and come back to this place where they belonged? Or would they continue to search in vain for him in the land that he had forsaken? Soulis Joseph would never know the answers to these questions. Nor would he ever see his father's rugged face or his mother's soft dark eyes again.

Trappers returned from St. George's Bay much later that summer with news of the merchant who had made Soulis Joseph his bond servant. The man had died a cruel death, as the consumption had indeed "consumed" his body. On his last days none of his white comrades would come near him, fearing the dreaded disease. He was carried to a shallow grave and covered in haste.

The two men who carried him to his coffinless grave – after they had first burned all of the papers found in the filthy hovel where he lived – were two Mi'kmaq. When they tossed the last shovelful of dirt into the hole, they had not only freed themselves, but also a young Mi'kmaq miles away.

Soulis Joseph stayed in the land of Ktaqamk and was accepted into the small tribe of Mi'kmaq settled into the deep reaches of Bay d'Espoir on the south coast of Newfoundland. He never returned to Nova Scotia or the Cape Breton Island area of his birth again.

He would come to love this land, which would shelter him and please him and give him children of his own. They would walk this same land for generations and carry with them the stories of an ancient way of life. They would never forget the best story of them all, though, the one about the old Mi'kmaq slave Soulis Joseph who had met a chained black slave aboard an English brig, and who had come to this land of freedom, this wonderful Ktaqamk.

II

Soulis Joe

There's gold, and it's haunting and haunting;
It's luring me as of old;
Yet it isn't the gold that I'm wanting;
So much as just finding the gold.
It's the great, big, broad land 'way up yonder,
It's the forests where silence has lease;
It's the beauty that thrills me with wonder,
It's the stillness that fills me with peace.

"The Spell of the Yukon"
Robert Service

13

"Soulis Joe was my great-grandfather, you know," Al told me one day as we munched on tuna sandwiches that he had made for us the night before. He was very proud of his Indian heritage.

"You know the name Soulis, fer sure. It's the name of the brook that crosses the Trans-Canada Highway just west of Benton, near Gander. Not only that, Soulis Brook flows from Soulis Pond. Makes sense, I guess," he said.

"Anyway, it was called after my great-grandfather, Soulis Joe. A lot of people calls it Benton Pond these days. Shouldn't be, though. The name of the old Mi'kmaq Indian, who must have trapped there years ago, should still be connected to that pond.

"There is another big pond that bears his name, too. Marked on the maps, even. Soulis Pond, just west of Conne River down on the south coast, a long ways from Benton. Travelled a lot, he did.

"Funny thing, though," Al continued, "we can find no evidence to prove that old Soulis Joe ever lived in the town of Benton. A bit strange, that is."

We talked more about Soulis Joe, about how ironic it was that the old Mi'kmaq had started a way of life for his progeny that would continue for several generations. Sitting there on the side of a raised embankment, under the shade of a birch tree, with the gentle wind rustling the green leaves above us, it was easy to see why Soulis Joe's bloodline continued to search for hidden mysteries. Prospecting was all about spending time in the out of doors.

"Best restaurant in the world," Al said, before tipping his ever-present bottle of blue liquid toward his mouth. I agreed with him and said so.

"Always a table by the window," I said, looking down over the deep wooded valley below and the sun's glistening on the stream that meandered through it.

"I could do with a little less of that pinkish sandwich spread, though," I remarked, looking at the mixture that Al had poured from a bottle onto our bread to hide the tuna.

Al grinned. "Better watch how you talks to the cook." He cooked almost all of our meals at camp and packed our day lunches. When we had fish, I was the one who washed and cooked it. Al liked the way I prepared cod.

We talked some more about the old Mi'kmaq native who had put Al and many others in his family on this long prospecting trail. We both agreed that even if there were no positive evidence that Al's great-grandfather, Soulis Joe, had lived in the town of Benton, the fact that his name appeared on government-issue maps meant that he had spent time in that area.

We talked like old friends, Al and I. For two men who had only recently met we found that we had much in common.

Especially evident was our love for the wilderness. It went beyond a mere liking to be "in the country." It was a feeling that we shared whenever we stepped away from the road and into the woods. It only took a few steps. Even with the sounds of traffic just behind us, we were as contented as if we were miles away, walking through some hidden valley.

* * *

He towered well over six feet, taller than most other Indians in the area; he was lanky and appeared a little gangly, though he was anything but. He was a blooded Mi'kmaq. His wiry frame hid beneath loose, ill-fitting clothing that made him appear skinny. Matty Mitchell didn't own much of a wardrobe.

His face was as weather-beaten as the hills he roamed. He was a quiet and unassuming man who kept to himself as much as possible. His lean frame was a well-muscled wonder. His face was angular, well-defined. The mouth above his prominent chin was full and crinkled at the corners below two cheek dimples, as if he smiled often. He had the smoke of a thousand campfires and the dust from a thousand trails in his eyes. A long, almost Grecian nose sloped away from a pair of deep-set eyes flanked on either side by crow's feet. His black hair peeked out under a hat that had long since served its usefulness. The skin of his face, as well as that of his long-fingered hands, was the rich, tawny colour that shirtless white men tried to draw from the burning sun.

He could have told countless stories of a way of life spent traversing the harsh country along Newfoundland's northwest coast. But usually he only spoke to his white bosses and workers, and only when asked a direct question. Then he would only give direct answers.

Matty could have answered every question about "the country" – what all Newfoundlanders call "out in the woods" – with added detail and information that few people of his time could manage. But he seldom did. Among his own kind, however, he raptured audiences with tales of adventure. His voice dripped that soft campfire blend of sound that was musical and haunting and always held a listener in a trance.

A dead muskrat hung from the rafters by a piece of hemp rope tied to its tail. "The muskrat is not only good fer fur, ya know," Matty Mitchell said as he stroked the animal's soft skin. Two young faces stared at him in awe.

Sitting in silence on the edge of a wooden stool against the wall, smoking the butt of a dirty cigarette, was another man who shared Matty's skin colour. He was as tall as Matty himself. His eyes were dark and intense. He listened as intently as did the two children. Soulis Joe was always a good listener; he loved to hear Matty Mitchell's stories.

"The water rat makes dem shiny white pearls. Dark ones, too," Matty said softly, his voice mesmerizing, believable.

"The white man loves 'em. I've seen dem lots of times. Mostly in brooks. Find dem rats an' you find dem pearls. Sure nuff, no rats, no fancy pearls."

He twirled the rope, spinning the animal as his blood-smeared knife traced a skilful arc through the dun-coloured skin, releasing the muskrat from its warm coat. Its pungent smell filled the shack. Neither Matty nor his audience seemed to notice.

Soulis Joe leaned back against the rough wall, tipping the wide stool a bit as he did so. He had heard Matty tell the story once before. It had earned the Mi'kmaq nothing but laughter in response.

Several white men had heard he knew where to find precious pearls and had asked him about it. The good-natured Matty told them that he did know where to find pearls: the muskrat made them. The loud laughter that followed left Matty bewildered. He couldn't understand the white men, who seemed to know nothing of the land they now called their own; when told about its many mysteries, they often responded with mockery.

The white men would have done well to listen to their betters. The muskrat did indeed play a large role in the growth of pearls inside these freshwater mollusks. Though Matty Mitchell never knew the reason why the clams only seemed to have pearls where muskrats frequented, he knew he was right.

The muskrat's favourite food is black-shelled freshwater mussels or clams. These kidney-shaped mollusks grow best in swift-flowing streams and shallow brooks. Whole beds of them can sometimes be found near ranging muskrats, by ponds or steady watercourses where the rodents build their hubbles.

The muskrat, like most rodents, carries a water-born parasite which clings to its rich fur. When the animal dives underwater to feed on the black freshwater shellfish, it twists and pivots and pulls its wiry body in a ferocious effort to tear the succulent morsels from their anchorage. This frantic harvesting exposes the foreign body, which is then scraped free from its host by the sharp edges of the clams.

Now begins one of the greatest mysteries of a world we are only beginning to understand. For inside the small shell, the organism begins its defence against the intruder. In an effort to rid itself of the foreigner, the mussel now starts the process of isolating the parasite within its own body. It secretes a nacreous fluid which hardens and develops around the parasite, destroying and holding prisoner the louse that has now become a pearl. Without even knowing why, Matty Mitchell had been right.

Soulis Joe knew what it was like to be laughed at. His people often related their stories to others who didn't want to know or even hear the truth, least of all from a native. But among their own kind, around trapline fires and between bare walls, the Mi'kmaq people never tired of the old tales.

Soulis Joe left Matty Mitchell's trapline shack later that day, and the next noon sun found him far away, heading south. He thought about the stories Matty told. The man was like a shaman among the Mi'kmaq. His fame, like his wisdom, grew as he aged. Soulis Joe wondered how the story of the muskrat could be true, but he didn't doubt that it was so.

Walking along a faint, unmarked trace, he ran a favourite tale of his own lineage over again in his head. His ancestor and his namesake, Amgwesewa'j – Black Soolie, the white men had called him – had come to this island even before the white man found the green hills he now walked over. It was said that he and many more of his people had come across the water from a distant land.

When some of the whites heard the story of how Black Soolie had not only seen a chained black-skinned man, but had even shared food with him aboard an English slaver, they laughed. No one around these parts had ever seen a black man. Niggers, the white men called them. Soulis Joe had never seen a black man either, but he never doubted that the man whose name he was given had seen one, and had fed him aboard the ship with the terrible smell, just like the elders had told him again and again.

Soulis Joe was leaving Conne River, or Miawpukek, for good. The memory of his courageous ancestor had made up his mind. He had been wanting to leave for close to a year now. He liked the people well enough; almost all of them were Mi'kmaq like himself.

Trapping was his way of life, and it took him many days distant from the settlement where he lived. West, up over the divide that separated the reach of the salt sea, and past the northwest end of yet another sea arm. The other trappers already referred to one of the large ponds where he had trapped and hunted for so long as Soulis Pond.

There were many other trappers in the town. None of them would think of trapping on another's line. When they came to one they walked on until they reached a different watershed. It was an unspoken rule among the Mi'kmaq people. But there were others, white trappers from the growing coastal villages, who didn't respect this unwritten law.

He was always looking for somewhere else to trap. It was this search that had taken him far enough away to meet with Matty Mitchell. Soulis Joe had no intention of taking up his trade in Matty's area. They had never even discussed it.

There was another reason why Soulis Joe had to leave Miawpukek for good. A priest had come and caused much trouble in the village. Priests didn't usually take up residence, but this one did.

He denounced their way of life and even their speech. He called them pagans, savages. The priest beat the children for talking in their own tongue. The ancient ceremonies were banned. Skin drums, which made the music they loved, were no longer allowed in their homes.

Their very culture, which had existed long before this white priest's Bible had even been written, was denounced. The hardest thing for Soulis Joe to accept was that some of his own people went along with the new priest's teachings. It was tearing the community apart.

Soulis Joe was not a steady churchgoer or a religious man, but he knew what he believed. He knew what was right and what was wrong. He also knew he couldn't win against a white man, and a priest, no less. So he had to leave.

In his wanderings he had heard about bays far to the east and north of Conne River that were rich in furs and still relatively untouched. Most of the white men in those bays were ardent fishermen who seldom trapped far from their coastal villages. There was one place in particular he wanted to visit. It was the bay the white men called Bonavista Bay. He knew that a few of his own kin had already settled and lived in peace in one of the reaches of that bay. Trapline tales told of a place with many rivers where the salmon always came, of a place where his people lived sheltered from the sea, yet close to the storehouse of riches from the open ocean.

14

When I interviewed the patriarch of the Keats family at his home in the town of Benton, he was ailing. He had lung disease. An oxygen tank sat nearby. Al introduced me to his father and his sister, Madeline, who was caring for him. The old man shook my hand with one that once held great reserves of strength. I could tell.

Theodore Keats was born in 1919 in Port Blandford, in the very bottom of Clode Sound. Uncle Ted, as he was known, had much to say about a pioneer's way of life. I took out my notebook at his kitchen table to write about his early years. I expected to hear about many hardships: never enough food, sparse, ragged clothes to wear, and *Your generation will never know what we went through, my son.* I was mistaken. Ted never talked about that at all.

"I was born in ruttin' season," he replied to my first question. "When the caribou are horny, and the moose, too. No moose to speak of then, though," he added with an easy laugh.

"October 20, the best time of the year for a huntin' man to be born. I been huntin' for one thing and another ever since."

I removed my jacket. The wood stove was giving off too much heat in the small room. When the old gentleman started to talk I was struck with his intensity. Like the true storyteller, who will always look his listener in the eye, his brown eyes held me. He spoke with such a passion and with complete detail – without being boring – that it was easy to see how the tales of his ancestors had been passed on.

"I never seen a hungry day, b'y," he replied when I asked him about the the old days.

"Lots of lean ones, you know, but I always found something for me gut and for my family's, too. Molasses for your bread, molasses for your tea. Not a thing wrong with it. Big thick slices of bread, too, I tell ya."

He held his wiry brown hands apart about ten inches.

"A workin' man will always find a way to provide. We never had much store-bought food, mind ya. Not many stores around then. Kept a man lean and well-muscled, ya know, luggin' logs on his back and walking a trapline and huntin'. Not a thing wrong with it. I'd do it all over again."

He came to Benton on the railway, he told me.

"Pumpin' the handle up and down on one of them old push cars. Wonderful things, you know they were. A man could cover a lot of country on one of them things. You needed strong arms to keep 'em going, though."

Ted was a railway section man. As the name suggests, he was responsible for a section of rail on both sides of the few houses in Benton at the time. He saw potential in the limitless forests all around the area and decided to make the town his home.

"There was another reason, too, you know," he told me. "That big pond out there" – he pointed out the window – "was called after my grandfather, Soulis Joe. No, I never did see the man. Wish I had, though. Must have been a smart man. Like his grandchildren, eh?" he laughed.

"He – Soulis Joe – never did live here in Benton, you know, even though the pond and the brook that flows from it carries his name. He lived for a time in Cannings Cove, over there in Bonavista South way. Died there, too. Poor feller, we can't even be sure of his gravesite. Sad, that is. A family should always know where their dead lie.

"Port Blandford, where I was born, is in the next bay over from Cannings Cove. We have relatives all over that area. All with Micmac blood in their veins. Mi'kmaq, they calls us now. Well, things should be called as they are. Got no problem with that." The pride of his heritage showed in his eyes.

* * *

Soulis Joe set out early one fine morning in mid-September. He carried a good knife, a sharp axe, a long bow with a dozen arrows, and a few snares. His packsack contained a packet of tea, some flour, a pan for boiling water and for cooking, and one battered tin mug. He would look for another place to live.

He was no different from countless generations of his people, always seeking new and better lands. And like the others of his race, he was also an adventurer. Although he was leaving his home for different reasons, he was looking forward to the prospect of walking new ground.

On the first day he travelled east along the Conne River itself. It was a way that was well-known to him; he walked along well-trodden trails on the south bank of the river. When he left the river valley and continued east, the trails faded and game became more plentiful. Soulis Joe saw signs of trappers along the banks of many streams that he crossed, and he walked on with no thought of stopping.

For two days the sun had been hiding behind a sullen cloud cover. It rained once for a couple of hours, and that night he slept under a hurried lean-to while his clothes dried over a small fire. He figured he was still travelling east, but with no sun to guide him he wasn't sure. He dozed on a bed of fir boughs, waking often to gaze at the dark sky.

Before he fell into a deep sleep, he saw what he was looking for in the sky. The clouds had cleared enough for him to see the firmament. He stood away from the firelight and stared upward. There it was, faint but sure, the North Star. His people had used it in their travels for centuries. This small star at the end of the small drinking gourd had always guided the knowing traveller north. He knew it never moved from its position. He was moving east.

Breaking a branch from an alder bush, Soulis Joe peeled a ring of bark from it so there would be no mistaking it among the debris that lay around the forest floor. Placing the stick on the ground away from his fire, he pointed it toward the North Star. He banked the fire with short ends of wood, then lay down inside his shelter and was fast asleep.

The morning sun was a glow over the treetops when he got up and stirred the ashes of his fire to make tea. Soulis Joe didn't have much of the precious leaf that he liked to smoke, but he put a little in the bottom of his pipe. He stared long at his night stick. He wanted to go east. It had been his intention long before he had set out on this journey. Still, the lands to the north had always drawn him. He didn't know why.

The sun broke in autumn splendour, dispelling the cool of night as if it had never been. Dew sparkled on every bush. Soulis Joe squinted toward the east.

He shrugged his pack over his broad shoulders and stepped away to the left of the burning sun. He had decided to delay his journey east for a while and walk northeast, where his spirit led him. The day started warm, summerlike. Indian summer, the white men called it. He knew this warm time came every year when the leaves turned. It could last for weeks. He was in no real hurry to be anywhere, was in fact already where he loved to be. The silence of the land held no fear for him. The ghosts of his kind who still haunted the deep valleys and hid among the night trees were what drew him on. Their mysterious presence was his strength. Knowing they had walked before him was his comfort. Maybe, if he could find the right place, he would trap all this fall and walk out toward the bay to the east during the winter.

The days stretched on. Wonderful days of discovery. Every valley different, every river flowing to him.

Then, late one evening, when he decided it was time to find a campsite for the coming night, he noticed something that puzzled him. The river he had just come upon was flowing the opposite way. It was moving in the same direction he was walking. Had he turned around? Was he so enjoying his trek over new country that he had somehow headed back the way he had come?

That night he laid down his night stick again and pointed it at the guiding star. The next morning showed that he hadn't turned around at all. He was still moving in a northeasterly direction. So was the river he had slept beside. He had heard about it before. Where the waters divided and flowed north instead of south was a turning point for many of the south coast trapping Indians.

Middle Ridge, some called it. A vast wilderness in the interior of the island where no man lived. Where some long-ago spirit had decided to turn the waters back from where they had sprung, to share their life-giving flow with the other side of this wondrous paradise. Soulis Joe walked on, following the waters.

Caribou were everywhere. Small herds and big. The yellowing boglands, as well as the naked barrens, were streaked with their ageless trails. Often he followed the paths for miles, the deer leads often going his way. At the mouth of every brook and in the coves of every pond and steady alike, ducks flew up at his approach. Soulis Joe snuck up on several and killed them with his arrows. He walked along as if he were the only human alive in a vast land.

In a few places he saw a scattered sign of trapping. He figured the trappers who had been here were Indian. Along the way, he trapped for meat, stopping long enough to half cure the very best of furs. Rabbits he killed without snares, using skilfully thrown rocks instead. With one arrow he killed a caribou that had not yet seen a full year of life. The meat was tender and easy to cook over his night fires.

One day he saw the spoor of a lone male wolf, its long oval prints distinct in the grey sand just above the water's edge. It was walking and had been there not long before. That same night he heard its call. It sounded sad and sorrowing, a wail born of the wilds, that pierced the night and keened through the shrouded forests, but got no answer.

15

Uncle Ted Keats didn't know why the spelling of Soulis Joe's name had changed, but he had his own theory on it.

"'Twas spelled every which way over the years," he said. "S-o-u-l-i-s, S-o-u-l-e-y, S-u-l-i-e, S-o-o-li-e, even S-e-w-l-i-s. We spells it S-u-l-e-y. What I think happened is this. My grandparents couldn't read nor write. Nor could my parents, as far as I know. So when it come time for one of their children to be christened or married, and the preacher asked them how they spelled their surname, what could they say?

"I figures the clergy recorded it down the way it sounded. What else could the man do?"

It sounded right to me.

Ted had started prospecting when he was twelve years old. Maybe less, he wasn't sure. He was looking for his grandfather's hidden silver mine.

"We grew up with the stories about Soulis Joe, how he was always looking for minerals along his trapline. We knew he found a silver showing that he never really revealed to anyone before he died. Not to be sure, anyway.

No one knows why. There were reports of places where he found gold, too. Somewhere around the Gander rivers, I think.

"We figured the wrong crowd tried to take advantage of him. He was too smart to let that happen."

They knew all about Matty Mitchell's discovery story, too, of course.

"Soulis Joe's find is still out there. We're pretty sure we know the area it's in. I'll never find it now, but my boys may. Who knows? Wouldn't that be something?"

Ted had stayed in Benton. He started a sawmill there – a push bench, he called it – and powered it with a John Deere tractor. He and his boys spent years chopping a living out of the nearby forests, not only for their own mill but for Bowaters, the pulp and paper company.

"I never stopped," he said. "If I wudn't cuttin' pit prop wood for Bowaters – they exported it overseas, you know – I was cuttin' logs for our own mill. I never let my boys stop, either." He chuckled. Al laughed with him.

Then they got into prospecting, and for Ted Keats and his family, life would never be the same.

"We got paid for walkin' around in the woods. Can you imagine?" He leaned back in his chair and laughed again. I could see where Al got his sense of humour.

They were employed by the huge mining company Noranda to do exploration work in Newfoundland. The Keatses walked thousands of miles, he told me, and then in later years were flown to remote areas. They stayed for days in tents and log cabins. They had a constant supply of food provided by the company, just the basics for the most part.

Always plenty, though. They were expected to augment their larder with game and fish they got themselves.

"It was a wonderful life," he said, a faraway look coming into his eyes. "Got paid for it good, too. Always a dollar when a man wanted one."

I asked him how much schooling he had.

"Not much schoolin'. Book One is all I got from the school in Port Blandford. Got a lot of education, though."

He told me he did geophysics work, extensive mapping, compass layouts, and used all the modern-day mineral-finding instruments of his day.

"I can use a good many of the ones they use now, too," he offered.

"You must be proud of yourself when you look back at all you and your family have done," I stated.

"Proud as a man can be without being vain. You know. I'm not that kind of man. More proud of my sons – and my daughters, too, as well as my grandchildren – than I am of myself. The Mi'kmaq ability to find things hidden will never die. You haven't heard the last of old Soulis Joe's bloodline."

I agreed with him. While he talked I wrote in my note-book, trying to capture his spirit.

Hung over all the walls were framed pictures of Ted and his children, and awards from mining companies and governments who appreciated the work this man had started.

I had one more question to ask Ted Keats. He sensed that the visit was just about over.

"Have a glass of rum with me," he said. It wasn't the usual *Would you like a drink?* It was more of a request to share a moment with him.

"I'd love to, Ted, b'y," I said. We were friends already.

Madeline placed a bottle on the table. Old Sam, black rum. Al poured his, mixed in a spoonful of sugar, and filled his glass with hot water from the kettle.

"The only way I likes the stuff is in a toddy," he said, a bit sheepishly.

Ted mixed his with "a drop o' cold water, to get the taste goin'." He approved when I mixed mine the same way.

"Black rum is made from molasses, you know. The same blackstrap molasses we used to put on our bread and stir in our tea. I thinks about it every time I haves a drink."

He raised his tumbler of rum shoulder high in salute, and when our three glasses clinked as one he said to me, "I appreciate what you are doing, you know." His eyes looked fully into mine. I would never receive a greater compliment.

Still, I had to ask that one last question.

"Are you a spiritual man?"

I'd thought about the question before and decided not to ask if he were a religious man. There is a big difference.

"Indeed I am," came the instant reply. "I've always believed in a Creator. I don't put much store in religion. Don't attend church very much, either. But here in my heart" – he placed a hand on his bony chest – "the only place where it really counts, I believe. I say three prayers every night before I turns in."

I could see that he was unashamed of his belief. I was curious about the three prayers and asked him if he would tell me what they were.

"The Lord's Prayer, of course, always that one first that every man knows. And then the Creed." Here he stopped. "And the other one."

When I looked up at him I figured I had gone too far. After all, a man's spiritual beliefs are private, and not all of us are willing to reveal what we feel.

He hesitated. "Well, I don't know." Then he said, "Maybe I will tell you. It's a prayer I heard time and again from my grandmother. I don't know if I can recite it aloud, though I says it every night. 'Tis not the same as thinking it, you know."

I agreed with him. I understood it well. Forming thoughts in my own head and then putting them to paper was sometimes challenging.

With a bit of encouragement from his son and daughter, he began. The lines rumbled up out of his throat. It was like a man revealing his soul.

"My God who makes the sun to know,
In blessed hours to rise;
He gives its light in esalow,
And sends it 'round the skies.
Much like the song in death fulfilled,
The labour of my day;
My time O Lord still marches on,
Along my heavenly way."

Allan and Madeline stared at their father with open mouths. They had never heard him say that prayer before. But it was more than that; they had never seen him reveal himself in such a way before. I sensed emotion in Ted, and thought his brown eyes were shinier than before.

I asked him how to spell the word "*esalow*." He didn't

know, he said. His grandmother couldn't read or write. But he believed it was a Mi'kmaq word.

Next, Ted drew my attention to the many photos along his walls. One in particular caught my eye. It was a colourful picture of a man sitting on a burro.

"Won't find many donkeys in these parts," he chuckled. "That's my son Fred looking for minerals down there in South America. Up in the mountains down there. Not bad for a feller from a small place like Benton, eh?"

I walked around the room studying this man's history, displayed at eye level for all welcomed visitors.

We left soon after. I shook Ted's hand as when I had entered. Putting on our boots in the porch, Al called, "See ya, Dad. I loves ya."

And from his father, out of sight in the kitchen: "I know you do, my son. And I loves you, too. You tell that Collins feller to come back, now."

16

One evening he watched several muskrats swim back and forth in the calm waters of a grass-covered cove that dug into the riverbank he was following. Here he made camp for the night.

After setting two traps for the tasty rodents, Soulis Joe thought about Matty Mitchell and his tale of pearls. Stepping out into the river shallows where the current was slow, he soon found what he was looking for: a bed of black clams, some of them the length of his hand. Reaching down, he pulled two handfuls of them from the clear water and headed to shore.

He had almost reached the sandy bottom near the steep rise, when he slipped on the slimy rocks underfoot. He went down, fully submerged in the cold water, suppressing a scream of pain that rose out of his gut. His right ankle was twisted, maybe even broken, he thought, so severe was the pain.

In misery he discarded the shellfish and crippled up out of the water before hobbling back to his campsite. He removed his boot and inspected his foot. He could move it without sharp pain. It wasn't broken, but it throbbed in pain with almost every beat of his heart. It was already swelling. He would have to do something before the foot swelled enough to prevent him from walking.

He went back to the river, where the water cooled his swollen ankle but did little to stop the pain. Gritting his teeth, Soulis Joe half walked, half crawled back to the area where he had set the muskrat traps. The grass growing tall out of the water was rooted in a bed of black mud. Gathering some mud he made his way back to his modest camp. He slathered his entire right foot with the muck and bound it with a soft layer of birch bark that he had gathered for his fire. He could do no more.

Soulis Joe didn't get much sleep that night. Morning found him stirring the ashes and coaxing life into the starving fire. He couldn't put any weight on his right foot; he would need a crutch. He had stumbled away from his camp, axe in hand, and was searching for the right branch to hold his weight when he thought he heard voices coming upstream. Soulis Joe folded his long frame into the bushes. Peering through the foliage, he waited. He was always wary of strangers on lonely inland trails.

The noises came closer, then stopped. Two men had seen his smouldering campfire as well as his shelter. Soulis Joe didn't breathe. His foot screamed for a change of position, but he didn't move a muscle.

Then a man's head appeared, the hair matted, black, and long. Then another. Both were staring at his campsite. They had their backs to him. One of them was carrying a long-barrelled gun. They appeared to be puzzled by the presence of the camp. Then, one of them turned his way, as if in realization that he was being watched. He was Mi'kmaq. He spoke to his companion, and the other turned toward the forest as well. This man was Indian, too.

Soulis Joe grunted as much with relief as with the pain. The two strangers heard him. The one with the gun pointed his firearm toward him. Soulis Joe called a greeting and tried to stand. The other Indian walked to the bushes, looked in, and offered his hand, which Soulis Joe accepted.

The two strangers were trappers from the north. They had been making trips into this part of the country for a couple of seasons now, taking furs from along the river they followed, but most times from brooks and streams that ran into the larger river. Soulis Joe told them he had seen their sign.

They talked until one of the men left, and after a while the others could hear him chopping in the woods. He soon returned with his hands filled with the stringy inner bark of the aspen tree, as well as some red bark from a young cherry tree which grew on the riverbank. These he mixed with water and held over the fire until they were soft and pliant. The Indian then removed them from the hot water and pounded them into a pinkish pulp. When it cooled he placed it over the dark, swollen bruise on Soulis Joe's foot. He encased the mixture again with the black mud from among the water grasses and bound it with birch bark, holding it all in place with thin strips of birch rind.

The three men talked for hours about traplines, river conditions, and fur-bearing animals. About caribou, black bears, and salmon and trout running in the rivers. About wolves they almost never saw anymore. They talked about relatives and meetings such as this one along the old trails.

They talked about the white men and their ways.

The two men lived close to the ocean, they told Soulis Joe, just inside the mouth of the big river they called Akilasiye'wa'kik – the Gander. It was a good place, they said. Lots of birds and sea fish. Geese came every spring and fall. Salmon tasted better when they were taken from their weirs in the river estuaries; their red flesh still held its sea fat.

The white men took salmon differently, they said. They used great brown meshes that in some places reached across the breadth of the river, taking the salmon in greater numbers than they could eat. The salmon they packed in hundreds of round wooden barrels layered with salt and bore them away in their large two-masted sailing ships to the lands of their kings far beyond the horizon. They came back for more again and again. They didn't seem to notice that the numbers of fish were already fewer.

The Indians told Soulis Joe the white men hunted caribou in like numbers, poling and pushing up the river in their boats and killing them with their long-barrelled rifles. Other times they had even seen the white men shoot dozens of the huge stags, before the tide took their boats back down the river with tangled antlers hanging over the gunnels. The rich meat was left behind.

The trio talked and shared stories and experiences, as wanderers do. Stories of the land and the endless trails is knowledge stored in the mind of the true wayfarer. Ever useful, always remembered.

When Soulis Joe questioned the trappers about the waterway he was following and how the waters had changed their flow days ago, they shared with him all the wilderness lore they knew of this vast area. The river he was following ran into the southwest end of the huge Gander Lake, which they called Akilasiye'wa'kik Qospem. There was another, larger river, they told him, more than a day west, which flowed out of the northwest end of the same lake. They were fast, powerful rivers. There were many brooks and streams that poured into the lake and kept the waters deep and very cold, even in the heat of summer.

They had never gone the entire length of the big lake, but as far as they knew there was only one river which emptied out of the lake. They didn't know of any trappers who worked the east end of the lake much, though everywhere else was trapped. When Soulis Joe told them of his desire to move to a bay far to the east, the trappers were proud to tell him that, not far from where he had seen the waters turn to the north, there were other rivers that flowed east. They had never followed their entire routes, but they had heard that they met with the ocean in Bonavista Bay.

The trappers informed him that this Bonavista Bay must be a huge one. Their elders, who had travelled far, had said that the east end of the big lake wasn't far from the north side of this same bay. This part of the bay also had many rivers where the salmon came. Soulis Joe listened, absorbing all the information the two men divulged. He already knew about the north side of this Bonavista Bay. He told the hunters there were Joes reported in that area as well; he believed two of his sisters could be there.

They talked long into the night. They laughed when Soulis Joe told them how he had wrenched his ankle, although it was short-lived fun. Every woodsman knows the danger of being injured in the wilderness. Especially when alone.

They knew about Matty Mitchell. His fame had spread through the wilderness along all of the hunting trails of his kinfolk. They hadn't known about the pearl-hiding clams, though. Laughing, they told Soulis Joe they would take more care with their footing when they searched the clam beds they knew of. They sometimes used freshwater clams for fish bait, not liking the stinky grey flesh of the mussels for their own use.

They shared their food and their stories, as is the way of all wilderness wanderers. Men who walked and explored new lands were always carriers of tales and deliverers of news at their rare meetings. It has always been so.

By the time the three men found their sleeping mats, they were friends. Soulis Joe's foot was no longer burning, though it was still swollen, and he slept, despite the snores coming from the other two men.

The next day, the two men took care of Soulis Joe's needs. They changed his bandage and added more of their medicine. By nightfall he could stand on his foot, and though his walk came with a limp, he was on his feet again.

They separated the next morning, the two trappers heading upstream and Soulis Joe downstream along the same river. They never looked back.

17

The search for hidden minerals by modern-day sourdoughs goes on all around us. The romantic, haunting lines of Robert Service may be forgotten to most, but "the men who moil for gold" are still very much with us. Only the methods of travel and exploration have changed.

The creeks still run clear into black rivers, bearing their hidden mysteries away from the hills as they flow. Away from the screams of highways and all the noise of our modern way of life, men and women still search for the "Klondike" that is always over the ridge and hidden deep in the next valley. Nature does not often relinquish her secrets willingly. It requires the spirit of the old forty-niners – still very much alive – to continue the search.

My father, Theo Collins, as well as my uncle Louis, both men in their eighties, told me several times about a vein of copper that can be seen out on the "Point of the Bay." The Point of the Bay is a long wooded ridge that forms the southern side of our harbour in Hare Bay. They went on to explain how their father had seen men drilling there with a driving hammer and steel when he was a young man, in the 1890s, making the exploration over a hundred years old. This mineralization was at a place called Picker's Point, on the western end of "The Point." My uncle Louis – who was an avid duck hunter in his younger years – explained further about the drill holes that he was sure were still there.

"My son, I've seen it not once but a dozen times. Not only that, but when the tide was full, making the water too deep to get around the point – Mineral Point, we calls it – I'd many times used the holes drilled out there fer support to steady me way along. You know what I mean, put me fingers in the drill holes and pull meself around. About as big as me arm, they are. Oh yes, they're there, all right. Unless someone has filled them in fer some reason."

He was very sincere, and though I had no reason to doubt him, I doubted that the holes could still be seen.

"Not only that, but we always heard the same vein of copper runs the full length of the Point of the Bay. They found traces on the eastern end as well, we always heard. Though to be honest I've never seen any signs of copper on that end of the Point."

Given the knowledge I had learned about finding minerals, I decided to have a look at this copper vein all the older people seemed to know about. Besides, this long and very rocky point of land was spread out in the path of the Dover fault.

The Dover fault, on the north side of Bonavista Bay, upon which the town of the same name now sits, is an enigma. It is the result of ancient collisions which saw entire continents forced upon each other's shores only to be wrenched apart again. The forces and the time span involved are beyond imagination. Still, the evidence is irrefutable.

The Dover fault was formed where the ancient continents of North Africa and Europe collided with America. Probably during the Jurassic age. The terrible up-thrusting stress of the fault forced the two immense land masses back to where they are today. The sea that filled the space between them was called the Lapetus Ocean. It was named for the Greek god Lapetus, father of Atlas, whose name is derived from the word Atlantic.

All of this happened four hundred million years ago, thank God. The resulting clash has long since cooled, leaving the volcanic slate and shale formations on the east or African side of the fault, and the multicoloured granite cliffs and boulders on its west and North American side.

The town of Dover sits on the centre of the dormant fault. Most of its houses are anchored to the mytonites – the locals call it iron rind – that may have been forced from either side of the active fault long ago. The old fault line rises out of the sea at the northeast end of the town and reaches all the way to Hermitage Bay. Here it again submerges its history beneath the Atlantic Ocean, staying under water all the way to the American seaboard, where it comes ashore again at the state of Maine.

The Point of the Bay, as everyone calls it, is a wooded peninsula which juts out of Freshwater Bay and into the sea on the southern side of Hare Bay. A few kilometres into the bay, its high southwestern end reaches for the town of Gambo, blocking a direct view of the ocean from that town, as well as helping to form a narrow channel through which the sea flushes the bay on every tide. The Point of the Bay is just one of the jagged vertebrae that help form the spine of the fault itself.

* * *

I left my truck on the road nearest the bill of the Point and started on foot one early morning. The radio said it would be a scorcher of a day with temperatures in the mid-thirties. I carried my big pack, which kept my full water bottles and light lunch just below the ambient temperature. I wore a thin shirt – no jacket – a baseball cap, sunglasses, cotton pants, and long rubber boots. In my hand I carried a long-handled prospector's hammer.

I had timed the beginning of my journey to correspond with the ebb tide. This would make my walking much easier; I knew there were several places along the route where cliffs jutted right up out of the sea. It was my hope that the low tide would allow me to walk around these edges rather than force me to take to the dense stands of tuckamore which grew to the very edge of the cliffs. I should have gone around when the tide was halfway out, to give me an extra couple of hours before it returned.

The first leg of the journey was easy. Leaving the last of the houses behind, I was soon crossing the long white "Nailers" beach. The southwest wind was soft, bringing the sweet smells of summer woods to mingle with ocean scents.

Long, beaded strings of kelp lay like a drenched green carpet, glistening and clinging with sticky red roots to the many rocks that stuck up out of the sand. Small tidal pools atop fissures in the rocks were clouded with teeny, many-legged, side-swimming creatures as well as trapped crabs that could have crawled out over the edges and away to the safety of the nearby water. Several gulls hunted along the landwash, flying along with downturned heads, waiting for whatever the retreating water would reveal.

As always, for me the old feeling of being at home whenever I was alone with all of nature's wonders swept over me, and I was as contented as though I were the first man to have ever walked these shores. Indeed, for me it was a fantastic first, for now I was equipped with new information. Every rock and cliff and jagged boulder drew my interest. All I had to do was stroll along, not stare at things too hard, try not to miss anything, and enjoy the hike, but most of all try to keep the voice of Allan Keats whispering in my ear as I went.

A small stream wandered out of the thick forest foliage and across the warm sand to mix with the slack tide. I knelt below the shaded bank and slaked my thirst, to conserve my own supply of water. I dunked my hat into the pool created where the brook fell to the beach, and placed it dripping on my head before moving on.

Heading out to the bill of the Point was a fascinating experience. Every turn in the shoreline presented another outcrop to explore. Rust oozed out of many of them – Al's voice said that was the first sign of minerals – as well as white layers of quartz.

Many of the cliffs carried their secrets away from the shore and buried them beneath the ocean. Still, there were more than enough mysteries in plain view for me to solve, if I could. I hammered away at the most promising showings.

You have to get below the outside rust if you can. All minerals will decay when exposed to the elements. You should always try to get below the weathering. Al's instructions were burned into my head.

I put solid pieces that held promise in plastic bags and marked GPS coordinates and topographic map bearings in my notebook. Then I deposited them into my pack, which, like the corpse of the old prospector Sam McGee, heavier and heavier grew.

I hadn't gone much farther before I regretted wearing the heavy boots. I longed for the lightweight freedom of my sneakers, and though the boots helped me around most of the ledges at water's edge, I longed to splash through the shallows, wet feet be damned. At several places along the route great veins of quartz appeared out of the forest floor above my head. I followed the outcrop of serrated lines all the way to the sea, where it disappeared into the blue surface. A couple more of these broken pieces found their way into my pack.

While I was selecting pieces of quartz and was about to discard one of them, something about it stopped me. There was something different about this piece, but what? I studied the white rock with my lens but could find nothing unusual about it. On my knees now, I selected another piece of about the same size – there were dozens to choose from – and found the answer. One of the pieces was much heavier than the other.

You have to be careful when looking at quartz. 'Tis usually white, but it can also be tinged blue or yellow. I've even seen it red. It can also be transparent or translucent. Sometimes it will smell bad. Stinky quartz, we calls it. It can sometimes be mistaken for barite, too, especially if the piece is overly heavy. Al's patient voice rang in my ear as if he were crouched on the shoreline with me.

I didn't know if the sample I was looking at was overly heavy or not; I just knew it was heavier than the other pieces around. I also knew, from my time spent in the oil fields in the Beaufort Sea, that barite was one of the main ingredients added to the mud in drill shafts when drilling for gas and oil. It was also used in the refining process of sugar. It is the sulfate of barium that is used in everything from paints to rat poison.

I had also forced down a disgusting meal of it a few years ago while undergoing medical radiology for a stomach problem. A barium swallow, they had called it.

By now I had reached the easternmost end of the Point and found not one visible sign of copper. Well, nothing that I was sure of. I figured the old stories about the copper showing were wrong. But I would go on as I had planned. I was enjoying the trip.

When I turned the point and headed southwest into the bay, I encountered the first of the sheer cliffs. These must be scaled over, walked around in waist-high water on the ocean side, or suffered through in the tangled black spruce growing out of the forest side. The decision was a quick one for me, for while the huge forests of the interior were easy to walk through, the crowded undergrowth along the coastline was an absolute nightmare. Sliding my packsack from my sweating shoulders, I removed my clothes down to my briefs. I stuffed my shirt and pants into the pack and, rolling down the calf-length rubber boots as close to my ankles as I could get, I stepped into the water.

After the first shock of the cold, I was elated. Now I had a much easier time of it. Walking in the water around the rocky shoreline gave me a better perspective of the cliffs protruding out of the woods. It also kept me cool. I figured the pack would keep my back from burning. The sun was overhead now and blazed out of a cloudless, windless sky. The heat was sweltering.

The ocean to my left was full of many memories for me. As a boy in the old "one lungers," I had passed along this shore many times, my father humming at the tiller while I crouched in the bow of the boat, with my chin jiggling on the stem to the throb of the engine. Later, too, in my own faster boats I had slipped alongside these cliffs, never knowing how difficult they would be to hike over.

I reached the place the locals called Beef Brook – why, I have no idea – and was looking forward to another drink of cool water, but the brook was almost dry. Only a meagre trickle of brown water seeped around rocks now overgrown with green algae. I refused to drink it. However, I did splash some of the water over my face as well as soak my hat. Drinking from one of my bottles, I moved on.

I walked over several shale or slate formations, banging at the rocks as I found them, to satisfy my curiosity. Granite boulders, some towering over my head, I paid little attention to, although I knew they could also contain gold or other metals.

I found mica easily enough. In one place the content of mica was so great I thought it might be schist, a metamorphic rock that always has a high concentration of mica. Then I figured it wasn't altered or layered enough for schist. I decided to take a piece of it back with me anyway. My pack wasn't light anymore.

The way in front of me was a stretch of straight shore, but very rugged. Jagged ends of cliffs stuck out of the forest and hurried across the narrow landwash to plunge into the Atlantic, presenting me with a formidable barrier. Most of them I walked around, in water that came up to my waist and dampened the bottom of my packsack. It was a hike of discovery for me. I couldn't believe I was getting paid for it.

Sediments – soil that had been dragged along by different forces ages ago and left behind on the earth's surface – were everywhere, as well as sulfides. Well, at least traces of them, in almost every boulder I banged at. All minerals are inorganic, and many form sulfides. Major ores, such as copper, lead, and silver, are sulfides. The telltale black or dark brown streaks are the result of their weathering.

Though I peered through my pince-nez as hard and as long as I could, I still found no evidence of the foretold copper. No pyrites, or even one sample of all the other "ites" that could be found in such rock formations as the ones I was travelling over.

The curved, pebbled beach of Father Ardin's Path End, better known as Uncle Joey Wells's Farm, was the next place of interest along my route. I descended the rocks on the north side of the beach, grateful for the easy walk along this stretch of shoreline. High above the sloping white beach, a once grassy field now grown over with raspberry bushes and trees still showed evidence of the old vegetable garden of its namesake. A stream of water worked its way through the white, sea-washed stones.

Here I sat and enjoyed my lunch. The gentle movement of the sea as it chuckled and tickled along the edge of the time-worn strand just beyond my dining table was soothing to a man's soul. Soaring high above me an eagle searched the water for a meal of his own. Closer he dropped, on wings that never moved, his distinctive white hijab moving from side to side, guiding the magnificent hunter along. The bird tilted its head once when it saw me and then wheeled away to seaward, still searching. Somewhere behind me, up in the trees, a raven croaked. I couldn't see him.

A steady drone of sound came to my ear almost the same time as a speeding white outboard boat appeared around the point, back the way I had walked. A shirtless man sat at the tiller. Two children sat on the middle seat ahead of him. The two kids wore bright red life jackets. The man at the tiller did not. The boat soon roared past.

Then the bay was quiet again, save for the soothing breeze's whispering through the canopy above my head and the swishing caress of the blue summer sea at my feet. I settled myself into the finer rocks above the waterline. They cooled my skin, and after a few minutes' work I had shrugged and squirmed and shaped a nest the full length of my body. Using my pack for a firm pillow – the rocks inside took some getting used to – I leaned back and decided to rest for just a few minutes.

18

Soulis Joe followed the east bank of the river. He had learned from the two trappers that the mouth of this river was very deep and difficult to cross. He was still in some pain from his ankle, but it was improving as he kept to the mossy banks just inside the treeline.

He didn't drink from the big river, but slaked his thirst from the cool streams that joined its flow. Late one evening by such a stream, after drinking his fill, he was bending again to fill his black kettle when he noticed a flash of yellow as the current moved the shallow water along.

Soulis Joe sat back on his haunches. He had lost sight of it now. Somewhere off to his left a loon trilled in the gathering dusk. It was flying, he knew by the sound. He turned his head, looking to the sky, and saw the bird through the trees. It was canting its wings for landing. Farther away and down low another loon called a greeting.

The lake must be near, he thought.

There it was again. No mistaking it. A small, yellowish object glinted in the water when the rays of the setting sun hit it. He leaned forward and tried to pick it up. His fingers dislodged the silt from the stream bed and his prize disappeared. He waited for the water to clear again.

There it was again, and next to it another, bigger piece. Bending again and working with his long fingers, he pulled both pieces of yellow rock out of the water and stepped back into the evening light to examine his find.

One of the pieces was about the size of the tip of his little finger. The golden colour that had drawn his attention was encased in a reddish-grey pebble. The smaller piece, the same golden colour, was on its own. It sparkled as he turned it over and over. He noticed something else. Unlike the other stones in the brook, this one had sharp edges. It couldn't have come far, he reasoned.

Searching all around the area, he found dozens of boulders in among the trees. Most of them were covered in yellow moss. All of them were composed of a coarse, dark green substance that almost looked as though they had been melted at one time. Some appeared to have grey crystal-like rocks embedded in their sides.

Soulis Joe turned his attention back to the brook, half expecting to find more of the gold-coloured rocks. He raked the bottom with a stick in a gentle sifting motion. A curtain of silt rose and washed away in the current, exposing the heavier pebbles beneath. He saw no more gold.

There was no doubt. It was gold. It was soft when he tried to bite it. He had heard somewhere that pure gold was soft and always showed the same golden colour no matter how it was turned. He wasn't excited at all. Although he knew of its value, he knew of no immediate use for the metal. He wondered where it had come from, how it had formed. Why was it here, standing out among the other stones of the same size? Was there more of it along this stream? It was late and he had to prepare for the night that was almost upon him. Tomorrow he would look further. Depositing the rare pebbles inside one of the deep pockets of his woollen breeks, Soulis Joe strode away.

The next day he picked his way upstream from the gold find. He could hear the sound of rushing water getting louder. The brook narrowed and the going became rough, slippery and very rocky. He stumbled once, his ankle still not as stable as it had been. He didn't know it then, but the stretched tendon would never fully heal.

Rounding a bend in the stream, he looked up ahead. The water narrowed to no more than two feet where it ran through a notch in a rust-coloured cliff. Here the water dropped over the cliff edge and plunged into a pool. When he reached the top of the dark pool he noticed on the left-hand side of the outcrop, just above the waterline and disappearing beneath the steady flow of water, a narrow vein that sparkled with miniscule grains of the same yellow metal he carried in his pocket.

Soulis Joe knelt and studied the find in his usual methodical way, trying to figure out what it was he was looking at. The cliff edge itself was sharp and no more than five or six feet in height on both sides of the brook. It was clear that the rock ran under the stream which had, over time, carved its way through and created its own channel. Although from a distance the entire rock appeared grey, it was in fact stained with a greenish, reddish, and sometimes yellowish patina that seemed to ooze from the rock. Or maybe it had been forced out, he thought.

The vertical vein flecked with scattered gold seemed to be trapped within the rock itself, and was no more than an inch wide and four feet long. Just below this four-foot vein was another, solid grey, mixed with a reddish colour, that ran at a right angle to the yellow one. Soulis Joe guessed that the vein below had acted as a dam and trapped the yellow one before it could go farther.

He banged at the blocked vein with the back of his axe, even though if he used the blade he could chop away more of the gold. No point in blunting his axe, he thought; he still had a long way to go and his axe was his most valuable tool. Still, he could see now where the gold he had found a hundred feet or so downstream probably came from. His curiosity was satisfied for now, the mystery solved, so he walked back down again.

Soulis Joe stopped when he reached the opening where the brook tumbled its way down over the bank to merge with the wide river. Staring upriver and down, and back in through the dark woods where he had found the gold nuggets, he put to memory every detail of his find. Then he turned and continued on his way and, as usual, never looked back.

* * *

Akilasiye'wa'kik Qospem was the biggest body of fresh water Soulis Joe had ever seen. The warm days continued, though the nights were getting colder. Huge stands of spruce and fir trees stretched as far up over the ridges as he could see. Majestic pine trees, standing so high he had to walk away from their massive trunks to see their tops, towered over the banks and beyond.

Groves of white birch, their leaves fading to yellow, swaying red maples, and scattered cherry and squashberry and dogberry trees grew in a mix of colours that was a joy to behold. The smells of autumn, too, that every true hunter knows and loves, were rising on the early morning mists.

The biting insects had all but gone from the forests. Long skeins of geese appeared over the treeline, their haunting cries echoing through the trees long before they came into view. Flocks of ducks sprang from every cove, some of them whistling as they flew. Dozens of pairs of loons, their families ready to fly behind them, swam away from the shoreline as Soulis Joe made his way along the south side of the huge blue lake.

In places where he had to leave the shoreline because of high cliffs and deep water, wooded glades and animal leads showed him the way. He filled his belly with rich-tasting blueberries. At one berry patch he startled two black bears that feasted on the sweet fruit. They disappeared in a few bounds, with only their humped backs showing above the bushes.

At one place on a promontory about halfway down the length of the lake, Soulis Joe spent some time sizing up the country, as he always did. He always studied his back trail too, knowing the way would look different when he came back. So far he had crossed several brooks running into the lake, but not one running out. The Mi'kmaq trapper could see on the other side of the lake and off to his left a wide swath in the forest canopy that seemed to go deep into the treeline.

It had to be the path of the river that flowed to the sea, the one the two trappers had told him about.

Still, from this vantage point it looked as though a major river might run away from the lake at the easternmost end as well. Not far up from the bottom of the lake and on the northeast side, a huge bowl dipped between two ridges. It could have been a huge bog, or maybe another body of water. He would go on.

19

A soft, rattling sound awakened me. The noise came from the waterline moving over the loose pebbles. It was several feet closer to me than it had been when I had lain down. The tide was full in the bay and had brought with it a stiff east wind. I had slept for more than an hour. I hurried my gear together, knowing the high water would make my trek much more difficult. However, I soon calmed down, realizing there wasn't anything I could do about the tide. I would just deal with my problems as I came to them. Tide, like time, sure enough waits for no man. Especially a dozing prospector.

Another hour of hiking found me turning west toward Picker's Point. Here the cliffs just burst out of the woods and dived into the ocean, allowing no passage, not even one narrow ledge for a man to walk around. Save for the sea the only way to continue was over them. I could have walked around some of them if I hadn't slept on the beach while the water deepened. I could even swim around the headlands, but not with a pack full of rock samples.

Donning my clothes again and draining my boots, I headed up over the cliffs. Back and away from the cliff edges I walked, the going easier as I penetrated deeper into the stunted black spruce and fir trees.

I made my way back to the shoreline several times, until I came to a place where I could once again walk in the water on the flooded landwash. It was still very hot despite the east wind, and after the woods hike I walked into the cold water at every opportunity.

I was now approaching the westernmost end of the Point of the Bay and was in the approximate area where I had been told the copper showing would be. From here I had a plain view of the houses in the town of Gambo at the western end of Freshwater Bay. It was so called because of the three large river systems that brought a constant supply of fresh water from deep in the island's interior, to merge with the sea.

I stared long into the bay, at the occasional car shimmering in the distance as it moved between the white houses on the streets of Gambo. This place must have looked quite different many thousands of years ago, I thought. According to geologists, before the last ice age, about thirteen thousand years ago or thereabouts, the waters of the mighty Gander River had flowed from the heart of our island and drained out here in this same Freshwater Bay.

It is believed by geologists that Gander Lake – which sits in another great fault, one that reaches in from the ocean at Notre Dame Bay and twists its way across Newfoundland all the way to the south coast, where it merges with the sea again at Hermitage Bay – was once a saltwater fjord. Seashells and fossils which predate the last ice age have been found, which proves that salt water had once covered the valley that now holds the fresh waters of Gander Lake.

After the immense glacier had receded, it had left behind a huge dam of sand and sediment, closing the sea forever from the basin that would become known as Gander Lake. The two wild rivers that found their way into this freshwater lake still needed an outlet to the sea. The ninety-seven-kilometre Northwest Gander and the seventy-seven-kilometre-long Southwest Gander soon raised the lake level until a release couldn't be denied, and the mighty Gander, one of Newfoundland's strongest rivers, sped away from the lake in search of the salt sea. It drained its 6,400 square kilometres of wet reserve northward, searching and twisting and carving its way through thirty-five kilometres of pristine wilderness, until it poured into the western end of Gander Bay. The great lake had found the ocean again and, because of its wanderings, had provided a conduit into the heart of a vast land which would be used by native peoples and others for hundreds of years to come.

It was believable, too, I thought – this theory about the Gander River valley – as I studied the hills behind which I knew Gander Lake lay. These forested hills that dammed the lake from this bay contained the richest soils and sands, while all around them were ridges of tumbled rocks. The lake itself stretched like a long flooded canyon, fifty-six kilometres from west to east, looking like it should drain itself into Freshwater Bay from its narrow east end. Instead, the Gander River leaves the lake from its northwest corner, just a few kilometres from where the two large incoming rivers merge.

I stepped away, my head filled with long-ago thoughts. I tried not only to figure out what had happened, but to accept it. It had to be, I decided. It made perfect sense, even to a layman like me. What a place to find minerals! Here I was, crossing the centre of the massive Dover fault, and on the edge of the fault line of the Gander river system. Better look carefully, I thought.

I saw nothing. I peered into every crevice of each and every cliff. I pulled bushes back from the high bank edges in my search, figuring that maybe the site had long since grown over. Nothing, only the shale and slate outcroppings, with their messy slides of shattered rock that were so common along this area of the Dover fault. My grandfather and grandmother Ford, who lived on the fault line in the town of Dover, had the same thin slabs of slate leading to their doorway. It made an excellent walk that my meticulous grandmother always swept clean.

I glanced back the way I had come, going over the route again and again in my mind. Had I missed something? If that were the case, I would need Al's experienced eye to find what had escaped mine. I waded around another of the many cliff edges, the bottom of my pack dragging in the water as I went.

I was crossing a narrow beach that over the years had somehow wedged itself between two jagged, slate-grey cliffs. Talus materials from the cliff face were everywhere, as if they had been forced out of the steep side of the ridge by a blast of dynamite.

Dynamite?

My thoughts brought a full awareness of what I was seeing. Canker streaks of green and yellowish brown marked not one, but three of the cliff outcrops. They looked like the great reaching claws of some immense green creature poised on the edge of the sea. I had found the copper showing!

20

The two trappers had heard correctly. There was no river running out of the east end of the lake, and the break between the ridges Soulis Joe had seen from so far away was another brook running into the lake. In all that distance there was no other water leaving this huge expanse.

He had followed this brook up from the northeast end of the lake one midday and was surprised to find another lake close to the bigger one. His instincts had been right about this opening between the hills.

He had seen no sign of trapping on any of the brooks he had passed over the last two days. The elusive lynx had left its round paw prints in many places for him to see, and once he caught a glimpse of the tawny cat's shadow. This was a hunter's land, for sure.

Twice more he found outcrops similar to the one where he had found the gold. One of them contained a vein of copper. The other was as rusty as an Englishman's knife. Probably iron, he thought, though there were traces of a shiny grey metal laced through the rock face.

He stepped to the edge of the trees, whose roots poured down over the very edge of the mouth of the brook. Before him and stretching away between two untouched and forested ridges was a serene lake. Hundreds of geese swam just out of reach of the brook's pull. Ducks swam among their larger cousins, quacking and squawking their annoyance. A splash of water demanded his attention, and another salmon flashed into the air right away, before falling sideways into the water. He could see schools of the fish swimming by, their presence betrayed by their dorsal fins breaking the surface.

Soulis Joe was pleased to see two beaver houses complete with their winter store of alders and branches, cut from large aspen trees the animals had gnawed down close to the water. One muskrat watched the noisy birds, its ears shifting as if wondering why they made so much racket. Out on the open pond he saw several loons. All along the shoreline were tracks of fox and beaver, and at least one bear track. A caribou lead came out of the woods, headed out into the brook, and continued up over the other side.

Instead of stepping out of his place of concealment, Soulis Joe stayed back among the trees until he reached a point on the shore farther along, where a shallow cove concealed him from the mouth of the busy brook. He saw no sign of any recent human activity. One axe mark on a large spruce tree that had long since healed itself was the only sign he could find that man had ever been here. He worked a piece of the healing gum from the bruised tree and placed it in his mouth.

As he prepared his night camp, the wind died and blackened the pond. The loons started their nightly serenade.

Maybe they are singing goodbye, he thought, before leaving as they always did, to go wherever it was they always went when leaves fell. An owl hooted from somewhere up in the trees behind him, and two more sounded farther along the shore. Salmon splashed at regular intervals. Another wedge of geese skirled across the sky, lowering and honking as they approached for their evening rest on the quiet pond.

Soulis Joe decided this place would do just fine. He had wandered far enough. For now.

* * *

In the days that followed, Soulis Joe felt more contented than he ever had in his life. He answered to no one and lived by his own wits and hands. He decided to construct a more permanent home for himself and chose a place sheltered among the heavy growth of trees away from the pond, but near enough to view the stretch of water.

The Mi'kmaq built a comfortable wigwam and framed it with straight young aspen trees that were easy to cut. He drove the sharpened ends of the poles through the moss cover and into the rich soil underneath, anchoring his home to the earth. The ribbed structure was no more than eight feet across at the bottom and angled vertically for ten feet or so. All of the poles met at the top, each lodged against the other and held in place by short branches.

There was no shortage of large white birch trees. Soulis Joe cut deep into the bark of one with his axe to remove the white covering as far up the tree as he could reach. He placed layers of the bark horizontally around the dwelling, starting at the bottom. Each layer overlapped the one beneath by several inches. Then he wedged more stout poles, free of knots, between the rib poles and over the spread bark, to hold the siding in place. A foot down from the top of the teepee he left open, to allow air to circulate and suck out the smoke from the fires he would burn in the centre of the camp.

On the day it was finished, Soulis Joe stood back in the dusky evening light, pleased with his accomplishment. The lean-to that he had built to shelter himself while he was building the wigwam, he left for other uses.

Now he could move into the more stable home. He had built it on a slope to ensure rain runoff. To further insulate himself from the heavy rains that were sure to come, with the back of his axe he dug a shallow moat around the higher side.

Inside the dwelling he stitched the soft tips of fir boughs, mixed with dried moss gathered from the forest floor, into a single wide bunk well above the earthen floor. The narrow single opening through which he had to crawl on his knees was covered for the time being with stitched layers of birch bark. He would replace it with the warm cured hide of a caribou as soon as he was able.

For two glorious weeks Soulis Joe hunted and trapped around the secluded ridges, sometimes staying out for overnight forays before returning to what he now referred to as home. He ate the tender flesh of roasted muskrat and the dark meat above the wide tail of the beaver. He flensed and cured their hides in the evenings that were growing darker and shorter.

Schools of salmon swam up the brook and disappeared into the black depths of the bigger water. He built small, rock-walled weirs along the shallows of the brook where the salmon had to pass, so he could trap the migrating fish and fling them ashore with his bare hands. This far up from the salt coast, the salmon that had not eaten since they left the ocean had lost much of their sea shine as well as their rich taste. But smoked over a low fire of dried alders and sweet-scented moss, they tasted all right to Soulis Joe.

One morning the cove was covered in ice that reached out from the shore for several hundred feet. Soulis Joe broke the thin layer of ice to fill his kettle before studying the sky for several minutes. Off to the east and rising up over the ridge, black clouds were thrusting their way inland. A damp breeze brushed across his hardened brown face.

After eating smoked meat and drinking the last of his tea, Soulis Joe left for the mouth of the brook. He kept his bow and short arrows ready and walked along the edge of the trees, skirting the dense alder beds that flourished above the banks of the brook mouth.

Swimming close to him and diving with their tail feathers jutting above the water were a dozen or more ducks. Just away from the pull of the brook, the edge of the ice from last night's frost was beginning to break away. Several flocks of geese swam along the ice rim and, except for the loud honks of one long-necked gander, remained quiet. He was close enough to see the different shades of grey and white bands below their jawlines. A huge raft of waterfowl gathered out over the band of ice and in the open pond. They too made little noise.

All this the Indian noticed and pondered over. A few of the nearby ducks dived again. When they did he loosed an arrow at the nearest duck, which tried to spring aloft. The arrow slid through its body and kept it on the water. The ducks next to it squawked a warning and flew away.

Another bird lifted its head from beneath the water and jumped to flight, and Soulis Joe wasted an arrow. The next one he didn't. The thin arrow pierced the breast of the duck before it had a chance to escape.

He waded out into the cold, knee-deep water and retrieved the floating arrow that had missed, before the tide could take it. Bending to retrieve the two ducks, he saw a cluster of black clams clinging to the bottom. The recent injury to his ankle and the pain that went along with it when he had last picked them came to mind. He smiled a bit at the memory and walked ashore with the two birds dangling from his right hand, the recovered arrow in his left.

All day long the threatening clouds filled and lowered the sky. Soulis Joe skinned the two ducks and stuffed the hearts and livers and the two gizzards inside their chest cavities. He had one of the birds skewered on a green alder over the fire inside his teepee when he heard the first heavy drops of rain patter against the walls.

That night the temperature dropped. The first snow of the season fell. The wet, plastering snow stuck to everything it touched and made branches droop with its sodden weight. The wind tore out of the north, shifted to the northeast, then blew even harder from the southeast. The temperature climbed with the southern air. The snow melted. Sometime during the night the wind shifted again. From the west this time, and by dawn most of the cloud cover had fled.

Stepping outside for the first time on this new day, Soulis

Joe watched the clouds racing away to the east. He had pretty much made up his mind the morning before, when he had seen the first ice in the cove. Now he was sure. He would leave as soon as he was packed.

He had trapped and cured only enough skins to carry in comfort on his back for a long distance. No more than thirty pounds, he judged. To this he added some dried meat, several fillets of smoked salmon, four cooked duck carcasses – no geese; he found them less tender than the ducks – about a gallon of mixed partridge- and blueberries, and a few white Indian Tea berries thrown among them for added flavour.

Standing in front of the teepee that had been his home for weeks, Soulis Joe took some time to look around before he left. The pond was free of ice again, the birds were still gathering, and the wind had calmed.

The wobbling cry of the loons had returned to the water. He knew this would not last. Winter was close at hand, and though this was a good place to spend the long winter months, he had other plans.

He would find his way to the deep reaches of the central part of the Bonavista Bay that he had heard so much about. This was a place for a man to live, he thought. A place to raise a family. Strong sons who would trap the waterways and walk the long evening ridges. Soft-skinned daughters to tend camp and care for more of his bloodline. He would be back when the swollen spring waters had run out of the country and the sun had warmed the land again.

21

I was as elated as if I had found gold nuggets scattered on the beach, free for the taking. In the same instant I knew the feeling that Al had told me about. It had nothing to do with money at all. It was the pure pleasure of finding one of nature's well-kept secrets. Almost like the triumphant feeling of figuring out a long-studied problem, or solving a mystery. I lifted my pack's strap from one shoulder, the weight taking the red bag to the ground with a thud. I staggered ahead, off balance. My load was about to get much heavier.

I explored Picker's Point for more than an hour, banging at the cliffs with my hammer, digging and scraping at the stained rock faces with much anticipation and sheer delight. It took great determination not to take dozens of samples with me. I reasoned that a few would do, and besides, my load was already too heavy. The shoreline would get much flatter now, I knew, and the pitted cliffs would give way to sandy coastline for the rest of my journey.

I had decided to leave my find and was hammering at the outermost edge of one of the outcrops when I saw it: the drill hole that my uncle had told me about, the one that I had doubted was there! Sure enough, there it was, about the size of my wrist and working its way into the cliff wall at about a twenty-degree angle.

It had been drilled by hand; the old prospectors used nothing more than a long steel drill and hammer and lots of corded muscle. I recalled something else that Uncle Louis had told me. His father, my grandfather, had told him that he knew nothing about the men or the name of the company that had drilled this site. However, he did remember that one of the workers was called "something Joe." My God! How could I have missed this possible connection? He hadn't said the worker was Joe something or other, but rather "something Joe," as if Joe was his last name.

I was so taken with my recollection that I sat down and stared long into the bay, wondering if it were possible. Could the man who had helped investigate this site have been Soulis Joe, the ancestor of Al's family? Maybe he was one of the other Joes of Mi'kmaq descent – this Joe family sure got around. Direct links of the family of Mi'kmaq Joes could be found everywhere, it seemed, including at the western end of Freshwater Bay.

The town names Middle Brook and Dark Cove have recently been assimilated into the one town name, Gambo. This Freshwater Bay, where I had been walking and searching all day, had gotten its name honestly enough, as had the name Middle Brook. Into this long bay, its east end hidden from the open sea by a narrow tickle divided by islands, flow three large rivers. At the southwestern end of the bay, the Gambo river pours out from deep in the interior. Farther down the bay, close to the town of Hare Bay, Traverse Brook finds the sea, its headwaters born from the ridges and marshlands that divide the waters from the Gander River far to the northeast. Middle Brook, as its name suggests, winds its way seaward between the other two. The three river systems bring great volumes of fresh water to the narrow bay.

Madeline Joe and Mary Jane Joe – who were almost certainly Soulis Joe's sisters – as well as other Mi'kmaq, lived at the Middle Brook site in 1876. It is believed by local historians that they had lived in Conne River but came to Middle Brook from Clode Sound. The same two communities that Soulis Joe had come from. Ted Keats had named his daughter Madeline, in honour of his grandfather's sister.

It is also believed that Sylvester Joe, who it is said accompanied William Cormack on the first European trek across the breadth of our island, was another relative of Soulis Joe's. In all likelihood it was the other way around; Cormack accompanied Sylvester Joe. However, there is no written proof of their blood kin.

Along Cormack's route of 1822, there were three high promontories, or "tolts," as the Mi'kmaq called them. One was the high blue hill of the Gander and Exploits rivers, or what is now called Mount Peyton. One was Hodges Hill of the mighty Exploits. The other and southernmost mountain Cormack called Mount Sylvester, thus crediting his indispensable Mi'kmaq guide forever in the annals of Newfoundland's exploration history.

Looking into this long, forested bay, with the sun leaving the sky as the water calmed and turned pink with the gloaming, it was easy to see why the Indians had chosen this place. It was sheltered from the open sea and only a couple of hours' paddle away from it. Waterfowl were abundant, with almost every cove and all of the estuaries of the three brooks hosting nesting birds. Seabirds came to the small islands near the shore, their rookeries a ready supply of protein-rich eggs. Migrating Canada geese flew into this freshwater valley early every spring and fall to feed and rest for their journey south.

Silvery fishes came from the sea every spring, summer, and fall. Caplin beyond counting came in huge shoals to the outer beaches. Bay seals and small pothead whales, as well as untold numbers of codfish, followed them to the same shores. The caplin rolled up over the sand with the tides, free for the taking. They deposited beds of spawn ankle deep above the waterline, where they rotted and stank, drawing black bears and foxes to the feast.

Schools of smelt and salmon abounded in the three rivers. The wide delta area of the shallow Gambo river was ideal for the fish weirs the Indian people built. The remains of these rock fish traps can still be seen to this day. Every pond had its abundance of trout. Caribou roamed the inland barrens. It must have been paradise.

Well, not quite. There are stories of conflict between the Indians and white men here in this quiet bay, as it was elsewhere when Indians and Europeans met. It is said the Indians stole salmon out of the white man's nets around shore, knives from their premises, and used their punts and rodneys without asking permission. They even cut the white man's nets when they set them across the rivers. Strange how all of this is known, yet no one knows why they are no longer here.

Just upstream from the mouth of the middle brook, on a

wooded bank on its north side, the town of Gambo has recognized and acknowledged its lost history. In from the swift brook, an Indian burial ground has finally been protected. The site has been fenced and cleared of trees. There are even symbols of grave mounds, and maybe a dozen or more bodies resting on this sacred knoll. They are without a doubt the last remains of a people who had called this bay home. Maybe even before the Europeans had taken what wasn't theirs to take. It would be sadder still to find buried somewhere else along one of these riverbanks the remains of another people, whose bones are painted a rusty red.

I took a notepad and pen from one of the pockets of my pack and started to write my thoughts. The sun had started its downward slope toward day's end, although the heat was still pressing against the land. I still had three or more kilometres to go. I'd have dark on my shoulder by the time I got back to my truck.

No matter now. I had found what I had come to see. I wrote how surprised I was to find what had been here for the seeing all this time. I also wondered onto the page if the spirit of old Soulis Joe was still here, and if it had been on a warm summer's day like this one when he had been here. Or maybe it was one of the other Joes.

Had he a good pack with energy foods to sustain him as he worked, and when he was tired at the end of his day, as I was, did he have a good home and hearty meal to go to? I doubted it. More than likely it was a rough tent he had pitched on one of the banks in this very cove, where he built a smoky fire to ward off the mosquitoes, before he sought his bough bunk for a much-needed rest.

The day breeze faded and died in the pinkish hue of dusk, until only small rivulets of waves whispered at my feet. For the moment the soothing sound was as the murmur of a forgotten people who had lived here long before my time. Where the ancient sea touched and carried away on its ebb the constant freshness of a new land, they too had lived and loved and only in long death received a belated respect.

With bulging packsack I staggered the last few steps back to my truck in the twilight of that evening. I was exhausted but very pleased. Looking out along the Point at the place where I had begun my day-long excursion, I couldn't believe that such an adventure had been waiting for years, just a few steps off the well-beaten path.

The copper showing on Mineral or Picker's Point – whatever you want to call it – is still there, of course, as is the drill hole. You have only to follow along Park Wood Drive on the south side of the town of Hare Bay, Bonavista Bay, until the pavement ends. Then follow the gravel road all the way to the curved beach with the lovely bay view.

Keep left until that road too ends. Then step through the tall grass and purple mountain iris onto the easy shoreline, and still keeping southwest, or left along the beach, follow the landwash out to the point. See if you can discover it for yourself.

22

Soulis Joe did return again to the lake that today bears his name. He enjoyed the solitude he always found in this area at the easternmost end of Gander Lake. However, he never stayed there or lived there on a permanent basis. It is not known why.

All that is known is that he trapped the area long enough for the other Mi'kmaq bands along the Gander River to refer to the lake and the lone brook flowing out of it as Soulis's own. He found the area he had heard so much about in the centre of Bonavista Bay and settled there. It is also known that he spent years trapping in Clode Sound in that bay. He walked alone, upstream and west along the rivers that emptied into the deep reaches there.

For years he trapped and explored south of where the waters divided, to the head of Bay d'Espoir and northwest to the Gander River area. Bay d'Espoir, the name the French explorers had stuck to the vast waterway that reaches inland from the rugged south coast of Newfoundland, meant "Bay of Hope." This far-reaching fjord is sandwiched between mysterious, high-walled mountains all along its length. To a sailor who is used to the open sea, this deep bay gives the feeling that he is sailing upon a huge freshwater lake, until the salt spray blows up over the sides of his vessel to sting his eyes and burn his throat with the assurance that he is still upon the ocean. The English pronounced it "Bay Despair."

It was on one of his many travels into this hinterland, not far from the mouth of one of the brooks that flow into Clode Sound, that Soulis Joe discovered another mineral element. It would make his name a legend in the area, almost as famous as Matty Mitchell's name was farther to the west.

Soulis Joe had been trapping earnestly for months and paying extra care to the curing of his pelts, in order to fetch top dollar for his furs. He was determined to buy an English rifle, a muzzleloading, long-barrelled flintlock. He had seen what these far-killing guns could do, and had even handled and used one once. But he wanted to own one.

The price of one of these new rifles could be as much as an entire season's trapping. Some of the merchants, he was told, would agree to trade one for furs piled as high as the full length of the gun. Soulis Joe walked all the way to St. George's Bay, where he had heard a chandler traded fairly with the Mi'kmaq.

By the time he mastered the accuracy of the weapon, the lead balls it required as ammunition had run out. He was counting the last of the precious round balls one evening when he remembered something. He had seen traces of the lead used to mould the heavy shot before. The metal was embedded in a small outcrop of rock just above the foundered ledge of a talus slope not far away. Directly below the ledge, a stream meandered through the forest floor before joining a larger brook beyond. He knew the place well. This was where he had found silver pebbles. He had passed it scores of times, most of the time giving it nothing more than a casual glance.

Soulis Joe had told a few people of his find. The stories about his "silver mine" grew and grew. Many wanted to know about it. All were curious. He had even been followed at least twice as he had set out on his trapline, but he lost his pursuers with ease in the trackless forest that he knew best. Once he led two treasure hunters for several miles, in a direction that had taken them far away from his find. He had sat in the bushes, undetected, as the two men hurried along toward what they were certain were untold riches. Soulis Joe smiled as they disappeared.

He set out, gun in hand, to see if he was right about the deposit of natural lead. The sun was already well above the treeline, and the early spring day promised to be warm by the time he had reached the midway point of the peninsula on the south side of the sound.

The rifle, which he carried in his hand – he seldom slung it over his shoulder – gave off a smell with the rising heat that was unique to Soulis Joe alone. Most of his younger life he had hunted using bow and arrow, and sometimes even spears. Using such weapons, it wasn't easy to get within killing range of the keen-nosed animals. He always crushed the heavy-scented needles of the white spruce between his big hands, smearing the musky, animal-like odour over his clothing, weapons, and traps alike.

When he started using the gun, he kept it well-oiled with a thin layer of bear fat, or sometimes with the greasy yellow skin of a goose. It reeked all the time, more so on hot days. Not satisfied with the pungent smell of the white spruce needles, and finding that he could never get the scent to penetrate the hard English oak of the gun's stock, he smeared and worked into the wood the oil from the scent glands found inside a male caribou's hind legs. After a few days, the smell faded to a faint animal scent that he ignored, but which permeated his space just the same. It was said that a blindfolded hunter could pick out Soulis Joe's gun from a stack of twenty by following his nose.

He journeyed back along his old trapline and then turned off onto an old trace that only he could follow. When he reached the stream where he had found his favourite silver-coloured rocks, he stopped. It was late; time to make camp. The warming earth brought the luscious scent of new growth to his nostrils. A soft wind rustled bright green leaves among the willows. Most of the trees were bursting with spring, eager to present their virgin foliage.

After making a night shelter and starting a campfire for his evening meal, he walked to a few white birch trees interspersed among the forest of black spruce. Glancing up he noticed that the tips of the big trees, though bursting with life, had not yet leafed. Using the point of his knife, he carved a hole into the birch's outer wood about an inch or so. Colourless sap started running out. Tasting it to make sure it was good to drink, he fashioned a funnel out of bark and inserted it in the hole. By the time he had done this, the sap was already dripping. He arranged a container to catch a quart or so of the nectar, then looked up the slope that ran down to the small brook.

A portion of the upper end of the outcrop had long ago

broken free and slid downward. The avalanche had created a wide scree line that tumbled down to a V point and entered the brook. Just below, where the talus flow entered the water, was where Soulis Joe had found the silver showing. He looked for it now, but the sun had left the evening sky and the pale light filtering through the forest shadowed the stream. He would wait for the light of full day.

The next morning, he couldn't find the piece of silver-bearing ore. He had thrown it back into the water in disgust more than two years ago, when he hadn't liked the attitude of at least one of the men who were seeking his "silver mine." It could have been washed downstream or settled deeper into the heavy silt of bed. No matter. He wouldn't waste time searching for it now. It wasn't what he had come for.

Climbing the hill toward the outcrop, he created several small avalanches of the brown and grey gravel under his feet. The rock face he was headed for was almost the same colour. To a casual passerby it would be nothing more than a light brown gravel deposit that had fallen from the cliff above. There were dozens of such places around. But to the keen-eyed Soulis Joe it was much more.

When he scraped away some of the loose top materials, he found the heavy rocks he was looking for. He picked up a fawn-coloured piece that was no bigger than his kettle but very heavy. He banged at it a couple of times with the back of his axe before it broke in two.

The fresh break revealed two shades of grey, one much duller than the other. The stone was shot through with flecks of the light grey mineral he knew as lead. There were other, brighter specks, too. Though smaller, when he turned the stone to the sun, they sparkled and shone like distant stars on a dark night. Silver.

Soulis Joe discarded the rock and covered it with the coarse gravel. This, too, was not what he was looking for. Though he had passed the little brook with the silver pebbles many times, he had only been to the rock face itself once before. He stood and waited, demanding his memory to reveal what he had stored there. And, just as simple as that, he saw it: a shady overhang that had survived the landslide, its bottom half pulverized and hidden along the slope he had just walked over. Above the overhang, trees flourished and reached away out of sight over the ridge.

Directly below the sheer cliff was where the slide had begun. Great grey boulders, mostly buried, poked up out of the loose gravel. Dead, half-buried trees, bleached white with countless burning suns, bared their leafless remains toward the cliff that had claimed them.

And then Soulis Joe found it. A couple of well-placed blows with his axe revealed the blue-grey galena he was after. The more he banged at the rock the more the grey seam opened. It was an inch wide in most places and as wide as six inches in a few. The vein ran for several feet to what he guessed was the northeast before it crawled under the brown and green moss of the rock face. He didn't know how much farther it went, nor did he care. He would take what he wanted for now and return when his supply ran out.

Using the blade of his axe, he took some of the soft lead from the widest part of the vein, enough to fill his shot bag twice, he figured. It would last awhile. He wasn't a man to waste his shot.

When he bent to cover the vein with gravel and debris, all around the edges he noticed sparkles of the silver he liked so much. They were no bigger than the gunpowder grains he carried inside his shirt to keep them dry. There were smaller lumps of dark yellow and deep black as well as long, twisted streaks of grey, all sure signs of tarnished silver. When Soulis Joe reached the bottom of the slide again he turned around, etching the scene in his memory.

The birch container he had used to catch the dripping birch sap was overflowing. Soulis Joe drank all that he could hold, loving the sweet taste of the clear liquid. He would gather more when he got closer to home.

His people had been drinking the nourishing sap from the white birch for hundreds of years. They never knew that the liquid they drank every spring contained fructose and glucose, vitamin C, potassium, calcium, phosphorous, and magnesium. It also contained manganese and sodium and was rich in iron. Everything the human body needed to function as it should was inside the white bark of this amazing tree. All they knew was that for a few weeks every spring this drink gave them extra energy, quenched their thirst better than water, and always made them feel good.

Soulis Joe was satisfied with his trip. He would be questioned again back at the village, once the others found out that he knew where to get his own lead. His dark eyes were already squinting with mirth. He loved the tricks he played on those who sought his secrets. Especially the white men.

23

The search for rare and common minerals alike goes on, of course. It is and always will be an essential part of our existence. We cannot live without them.

Confined boardrooms and offices are where Kevin Keats now spends most of his time. They are just as important to mineral exploration as the endless out of doors, where his father spends his. In today's economy-based world, the man in the stuffy office has more say in the actual looking for minerals than does the prospector who finds them. This fact was never so obvious to me as during the writing of this book.

Man's insatiable appetite for metals has not slowed since the Bronze Age. This hardened alloy of copper and tin, which has been dated as far back as 3500 BC in the Middle East, wasn't recorded in our part of the world until around 1,000 years ago. Historians dispute its origin in North America. Some say it was introduced by the Europeans. Others are sure our own natives discovered it themselves, by natural invention.

Even though the Egyptians were known to wear beads of polished iron in 4000 BC, they were of meteoric origin and not mined. We know the first people to use iron were the Hittites of Asia Minor. For years they had managed to keep secret their method of hammering, heating, and then dipping the metal into cold water to harden it. Their weapons were always superior to their enemy's. The Hittite empire fell just after 1200 BC, and their secret was out.

Iron, the most common of all the earth's metals, and the mother of its alloy, steel, have been the basis for almost all of our inventions ever since. Ribbons of track, long enough to wrap our planet in dozens of steel hoops, transport us ever faster along our path of industry.

Precious metals, like gold and silver, and even diamonds, which sometimes escape the human eye, are mined. Modern devices sift through mountains of tailings, extracting countless minute particles of hidden wealth.

"What does gold look like?" I asked Al one day.

"Everyone asks the same question," he said. "The answer is always the same. Gold is gold. In the ground or worn on your ring finger, it looks pretty much the same. I see you don't wear a ring, though," he added, glancing at my left hand.

"No, Al, b'y, I can't stand the feel of 'em on my fingers. If I did wear one, though, it would be silver. It's my favourite colour."

"I don't have much use for rings myself, as you can see," was his reply.

He was standing calf deep in a brook several miles south of the Southwest Gander River. He was about to show me the secret to panning for gold.

"This searching for metals and such is really like looking for our own roots, you know," he said, unhooking the bungee cords from the quad.

"How do you figure that?"

"Well, I've read somewhere that there are seven minerals in our bodies, without which we would die. Most of 'em the same ones you and I are searching for. If that is the case, and if we came from the earth and are going to return to it, like the Good Book says, then aren't we trying to find our own selves?"

Deep stuff, I thought. I told him so.

The seven minerals Al was talking about were indeed present in all humans: calcium, iodine, magnesium, phosphorus, selenium, and even the hard minerals like zinc and iron. We couldn't function without them. Four per cent of our body mass consists of minerals. We also need trace amounts of chromium, copper, fluoride, manganese, molybdenum, chloride, potassium, and sodium.

I knew all about the tools needed for such a gold search, which were strapped above the running board of his quad. There were three contributing pieces involved in Al's way of doing it. I always thought they would rattle like a peddler's pack when carried over miles of rough terrain on a bouncing ATV. I never heard them make a sound.

"Did you know this is Sunday morning?" he asked.

I handed him the panning gear and went back for the shovel.

"You know what else? We're in church. Just listen to the singing."

We were both still then for several minutes. The forest

music was in surround sound. Dozens of songbirds flitted in high tree and low bush, singing as they moved. A soft summer breeze, as melodious as any zephyr, rustled the leaves and the white hair on Al's head and sighed away through the wind section. The tumble of the running brook provided the background harmony, blending and swelling the morning choir.

"Service is over. Time for the collection," Al chuckled.

He scraped the nose of his rubber boot over the brook's bottom, dislodging the silt. Then he brushed a few blackflies away from his neck – he ignored the hordes of hovering mosquitoes – and placed the panning gear below the water, into the gravel bowl he had made with his foot. Two of the pieces weren't pans at all, but aluminum screens about twenty inches in diameter. One had a coarse mesh, the other one fine. The third and bottom piece was the gathering pan. It was a solid bluish metal that had been heated to burn off even the minutest traces of oil from human hands. Al explained to me that gold flakes could stick to a faint film of body oil and float away during the final panning process.

The three pans settled on the sandy bottom as he wanted them to. I filled the top one with several shovelfuls of red, brown, and grey sand. Taking only the top pan, my mentor went to work.

He began with a vigorous swirling motion, lifting and dipping the pan in and out of the running water. Like magic, only the larger and heavier stones remained in the top pan. Discarding it and its contents, he reached below for the second pan, which had caught the residue from the first one. He repeated the same shaking, shifting, shuffling motion with this finer-meshed pan. The silt from the operation floated away downstream. Now all that was left was the last and true pan holding the separated sediments.

This third and most important pan was shaped like a Chinese coolie hat, its concave sides sliding down to a depressed nipple in its centre. Now Al's practiced skill became more evident. Though it looked easy, I was quick to learn that it wasn't. He swirled the contents around and around, allowing the water to flush the "tailings" out over the pan's lip with each controlled dip. Soon, all that remained of the mound of gravel he had started with was a handful of "fines."

With a final swirl of both wrists, Al tipped the pan half over, allowing all of the water and most of the gravel to trail up from the bottom, and lie in a narrow string the length of the pan's surface. Only the "heavies" remained. We hoped to find gold, the heaviest of all, in the "nipple" at the bottom.

Holding the pan at the desired angle, he bent over and used the powerful lens dangling from his neck.

"Garnets," he said, as if to himself. "A few tiny red ones, a nice bit o' quartz, and – oh, there it is – one, two flecks of gold. Hasn't travelled far, either, by the looks of it."

I stared long at my first natural gold showing that Sunday morning, feeling as blessed as any Bible-thumping preacher. We found many more such traces during the coming weeks, but the day when that first one glinted at me by the side of that bubbling brook will always remain in my memory as the Sunday of Gold.

* * *

They have found many treasures, this family of Soulis Joe's. They have produced mines that have generated millions of dollars into the economy, and potential finds that only await economy stability before they burst into public knowledge. There are no less than twenty-one members of the Keats family involved in the prospecting-mining industry.

In the end, Soulis Joe's real treasure isn't lost at all. It is still carried along the old trails where his spirit will forever search. It is in the new and modern finds brought away from the secret hills on the eager young backs of his bloodline.

I knew this spirit would never die when Al announced to me one day, after forty years or so of prospecting, that he would like to bang rocks on the surface of the moon, or, as I put it, "To boldly go were no man has banged rocks before."

In Kevin's youth, Al took him on some of his prospecting forays and carried him in his prospector's pack when the boy's little legs gave out. And when the same young boy played prospecting games instead of cowboy ones, searching without knowing, he carried the spirit of his Mi'kmaq ancestor.

It is more than the keen-eyed ability to find rare minerals that others pass. It is the age-old quest for new and better things, like that which brought the slave Black Soolie from the tyranny of an uncaring foreign people who had claimed his homeland, his beliefs, his inheritance. In his search for family and a new life, he found peace and awareness instead.

And home.

As long as mankind is in need, there will always be those

who seek. The quest for knowledge and survival and hidden mysteries demands that the lone adventurer, hammer in hand, follow along Soulis Joe's back trail.

It is my hope that his family will always find his spirit, for therein lies his treasure.

* * *

The legend of Soulis Joe's silver mine is still alive today. It also appears to be just as lost. The mystery only deepens with age and the telling of it. Some say the old Indian told a merchant who didn't show any interest. Others say he waited too long before he revealed his find, and by that time he was too old to lead a party into the site. A few say he tried to describe the site as he grew older, but searchers could never find it. Some believe he intentionally led them astray, up to his old tricks.

Some don't believe at all, but they are not of Soulis Joe's blood. His bloodline is sure it is still there in some secluded valley. Or closer than they thought, on the side of some forested ridge. It may only require the dislodging of a boot's length of slippery summer moss. Or for a knowing eye to look beneath the roots of a windblown tree and there see the telling signs of the old hunter's treasure. Who knows?

This we know. Soulis Joe met and fell in love with a young woman in what the English colonists called Cannings Cove, and on a clear December 19 evening in 1881, Soulis Julian Joe married Rebecca Wells.

Today, somewhere in Cannings Cove, out on a long,

lonely point jutting into the ocean, or on some high bank of a sheltered cove, the old hunter, trapper, explorer, adventurer, and one of this island of Newfoundland's first prospectors, lies in an unknown, unmarked grave.

24

Soulis Joe paused for a much-needed smoke, fishing the tobacco from the back pocket of his well-worn breeks. Inhaling deeply, he set his lean frame back on a boulder hard by the faint trail. Behind him he could hear the other travellers on the trail.

Noisy buggers, he thought, spitting into the low bushes. *Good thing we're not huntin'*.

The noise from the path increased and a man soon appeared, not in the trail at all but pushing apart the dense alders that lined the trace. Seeing Soulis Joe sitting and enjoying a smoke in comfort, the straggler making all the racket crashed through the remaining tangled growth with an angry curse and stumbled into the clearing near the Indian.

"Why the hell did you leave us back there?" he demanded. He was sweating profusely, his fat jowls spattered with red blotches of heat. "We are paying you to guide us, and by gawd that's what you'll do!"

By this time the second man had managed to find them, coming just as far off the trail and making twice as much noise. Unlike his overweight partner, he was tall and scrawny – the complete opposite of the other, in fact – and appeared a bit clumsy. He seemed to be more frightened of being alone in the woods than anything else and, seeing the other two men, he stumbled into their midst. He bent down, placed his hands on his knees, and took a deep breath.

"Where is this place you told us about?" the big one demanded, removing a packsack from his back, the straps of which had slid down his narrow shoulders, making him look even more dishevelled. "We won't find it if you keep running away from us."

"Jest figured any man could follow a trail," Soulis Joe said. "I been waitin' fer you. We are almost there."

Following the example of the first man, who appeared to be in charge, the tall fellow also loosed a small pack and let it slide to the ground. This pack, like the one the other carried, appeared to contain very little of anything. Just for show, Soulis Joe figured.

"Time for lunch," the big man announced. "We've been walking for hours and I'm starved." And without waiting to see if the others agreed with his decision, he slumped to the ground, the air expelling from his open mouth as he did so. He groaned with the effort of lowering his weight.

Knowing it would be futile to tell the man they had only been walking for an hour, Soulis Joe tried something else. "No water here fer boil-up. Jest a little bit along there's a fine brook with good water."

"You should have told us that before you left us. The hell with a boil-up. What is it with you Newfs" – he made it sound like a dirty word – "always wanting to boil the kettle? Pass me a couple of tins of stuff from your pack."

Soulis Joe removed the heavy bag from his back and staggered forward a little when the weight left his shoulders. Placing the grub bag before the hungry man, he once again sat on the rock, running a hand through his black hair, watching.

The fat man fumbled with the leather straps and managed to get the pack open. He dumped most of its contents on the ground. Selecting a large tin of beans with pork from the tumbled pile, he sat back on his thick haunches and demanded a can opener from their Indian guide. Bending down, Soulis Joe dug into one of the side pockets of the pack and produced a spoon. He grasped the tin in his left hand, held the spoon vertically in his right hand, then pressed the spoon edge against the lip of the can in a sawing motion. The sweating man stared at him as he sliced open the container. Soulis Joe strode toward the man, stuffed the spoon deep into the tin, and placed it before him.

"Well, by gawd, did you ever see the like?" exclaimed the tall fellow. "He opened the beans with his spoon! Never seen that done before." He selected a can for himself from the pack's spilled contents and handed it to Soulis Joe.

"Ever hear talk of a can opener in this neck of the woods?" the fat guy drawled. "Or at least use that knife you carry on your belt. Why the spoon? Trying to impress us?"

"I didn't bring a can opener; I use a spoon. The metal from the can dulls the knife edge," Soulis Joe replied in a reasonable voice. He opened the second can for the tall guy, who watched the procedure as if it were some major breakthrough. Soulis Joe felt like he was talking to children. He didn't bother to tell the man that he didn't own a can opener and didn't have much use for one anyway. He wasn't fond of canned foods, and besides, he had no money to purchase food from a store.

"This place where you *say* you saw the silver rocks. How much farther is it, or have you lost your way?"

Soulis Joe's eyes narrowed to a thin crease, but when he spoke his voice was level, controlled.

"I never told you about any silver rocks. I said I knew where a stream showed some sparkles of colour. Kind of silver and glinting, but kind of dull, too. It's not far from here."

What Soulis Joe didn't tell the man was that, if he had gone along the trail, which for him was easy to follow, he would have had the kettle boiled by the stream just ahead and had a hot cup of tea prepared for each of them by the time the greenhorns had caught up with him.

"Changing your tune already, are you, Chief? Not turning Injun on me, are you?" The big one roared with laughter that shook his belly. Half-chewed beans peeked out of his mouth.

"Ah, leave him alone, Foster, the man is doing all right. Besides, I never heard him say anything about silver rocks either. Traces in a stream is what we came to see."

Soulis Joe glanced at the tall one in surprise. *He don't say much*, he thought, *but when he does he's okay*. However, he would like to "lose" the one with the big gut – and mouth to match – out in these woods for a night or so alone.

Foster glared a warning at the speaker, who looked uncomfortable. The two men talked some among themselves. For the most part they talked as if Soulis Joe weren't even there. Listening, he learned the tall one's name was Josh.

They had not introduced themselves to him back at Port Blandford, where they had started from this morning. They had been late starting, much to Soulis Joe's disgust, because he had to wait for the two to get out of bed and work through a big breakfast at the only boarding house in the village. Soulis Joe, who had arisen early, as was his nature, had slept until the first light of dawn had crept through the open window above the kitchen daybed at a friend's house.

The day before, he had walked from his home village of Cannings Cove, a few miles to the northeast of Port Blandford, to guide these men to his "secret" find. After leaving the community, and walking along the lengthy salt-water indraft to the mouth of a narrow brook which babbled over a rocky bed into a shallow estuary, he had endured a constant barrage of complaints and abuse behind him all the way along, all of it from the bigger of the two companions.

When they started again, the two men had their bellies full of canned food. Soulis Joe hadn't eaten at all, nor had he been asked to share their provisions. He kept just ahead of them without further incident, but with constant grumbling coming from Foster in the rear.

They arrived at a small stream. It was a pleasant place. Low bushes grew to the very edge of knee-high banks of rock, below which the shallow stream chuckled along only a few inches above a colourful pebbly bottom.

"Is this the place? I don't see any silver," Foster asked in an excited voice, removing his pack.

Soulis Joe knew now why the two had brought the light packs. The fools expected to carry out a load of silver. Foster was glaring down into the brook, his displeasure evident. He was not a patient man. Greed showed in his face as he turned and started to lash out at Soulis Joe again. Josh tried to reason with him once more, but it was a losing battle. Soulis Joe could see that Foster was in charge and that Josh was his underling. Everyone was an underling to this man, he thought. He would put an end to it.

"Last year, in the fall, I come along here, tending my trapline. We had a lot of snow this winter. A big spring runoff. I'd say this brook flooded many times its size. Overflowed its banks, maybe. The rushing water in springtime moves rocks along the bottom. Everything changes. The silver rocks washed away, maybe. May never be seen again. Sorry about that."

"You mean you took us all the way out here in this bloody wilderness for nothing? By Christ, you won't be getting the two dollars for your so-called services."

Soulis Joe had had enough of the fat man's insults.

"I'm going to check a little farther upstream jest to make sure," he said.

And without waiting for the others to join him or dispute his decision, he vanished without a sound into the forest that edged the clearing. He could hear Foster cursing and shouting behind him as he moved deeper into the woods. Unmindful, he hurried on until the sounds of the men faded away in the gentle silence of the forest. Following the twisted leads through a tall stand of black spruce, he strode away among the shadows.

He hadn't gone far before he came to another tiny stream. It was a tributary that ran into the same brook where he had left Josh and Foster. He knelt down and felt along the rocky bottom, looking for what he knew was there. He saw nothing at first. Looking too hard, he decided. He glanced away from the water and relaxed for a bit before he peered again into the glassy, slow-moving water. There it was, just as he knew it would be. Several brown-grey stones flecked with silver, his favourite colour, showed in the narrow trickle, and poked out from beneath the undercut of the red gravel bank opposite. He couldn't be sure, but he thought there were a few more of them than the last time he had come through here.

Satisfied, Soulis Joe reached into the water and picked up one of the fist-sized rocks. He turned it over several times, just to see the sunlight bounce off the flecks of silver metal within. From the stream he selected another shiny rock, smaller than the first one, and jagged, but showing the same silver traces and specks of another, duller metal. Lead.

Tossing the bigger rock back into the brook, he placed the second one deep into one of the pockets of his woollen jacket. Stepping into the shallow water, Soulis Joe scuffed his feet all along the bottom, stirring the rocky bed and mixing the coarse soil to bury the glinting pebbles.

He stepped out of the stream and stood up straight, watching the brook cleanse itself of the cloud of silt he had made. A few minutes more and the water was as clean as before. Now, stare as he might, he could see no sign of the silver ore. He looked away for a while and then looked back into the stream again. He saw nothing, only the narrow watercourse running off to the larger brook.

The Mi'kmaq looked around once, in all directions, and without a backward glance walked back the way he had come. *Maybe I'll come back one day*, he thought, *and show my find to someone*. It wouldn't be revealed to the two men with him today, of that he was sure.

Stepping back into the glade where the others were still scouring the stream, he emerged just as silently as he had disappeared. Soulis Joe was reaching for the heavy pack before the two men saw him. Foster started to rise from his knees in the riverbed. He was soaked. His mouth contorted for the tirade he had prepared for Soulis Joe, but his guide didn't give him the chance.

"Nothing here to be seen. The bottom of the brook is changed, like I said. Dark's on the way. We'd better get going."

"We're not leaving yet, you Indian bastard!" the fat man railed. "You got to find that silver for us!"

But Soulis Joe was already shrugging the big pack into a comfortable position on his back and running his fingers under the wide straps where they bit into his shoulder. Drawing himself to his full height, well above the angry Foster, he answered. His voice was deep, no-nonsense. His black eyes flashed, holding all of his anger within.

"Find your own way along, then. I'm leavin'."

And he just walked away.

Yelling toward their Mi'kmaq guide and without taking time to collect their own packs, the two greenhorns hurried after Soulis Joe. All they saw was his determined back, and not the broad smile that creased his black eyes and wrinkled his rugged face.

Epilogue

She sat before a large, gilt-rimmed mirror and leaned back against the rare Chippendale chair.

"What to wear?" she pondered. "Silver or gold?"

Her black evening gown clung to her svelte body. It dipped at the front to show the swell of her cleavage. Bending forward, she opened the ornate mahogany jewellery box, half the length of her antique bureau.

Pulling a looping lace of gold chain from one of the intricate chests within the velvet-lined container, she admired the sparkle of diamonds embedded in the pendant dangling at the end of the lush string, before placing it over her dark, coiffured hair. She draped it down over her chest and gave a long sigh as the cool metal settled between the warmth of her full breasts.

There was a faint noise behind her as the door opened. She glanced up and watched her husband walk across the room to the mirror. He approached, attempting to fasten a diagonal gold link through his shirt's crisp cuff.

"Damn things. I can never get them hooked," he exclaimed.

He sounded frustrated. She reached up to help, with slender fingers tipped with long, shining red nails that matched her lip gloss. She finished the task without comment. Looking down at the image of his wife in the polished glass, he broke the silence between them.

"I prefer silver with black, dear. I wish you'd remember these things." His annoyance was plain in his voice. "Downstairs in ten minutes. Don't be late," he finished.

Without looking up, she heard him cross the carpeted room. She caught a glance of his broad back in the mirror just before he shut the door on his way out, loudly, a sure sign of his displeasure. But she was never late, and he knew it. She didn't want to go to that evening's event anyway. She didn't believe her husband wanted to be there either, but these ceremonies were a necessary evil if he wanted to keep on top of the precious-metals scene.

They owned three high-end jewellery boutiques in the city of Toronto and were in the process of expanding to Montreal. The French loved fine jewellery. She sensed he regretted the decision he had made years ago to include her in the ownership of the business.

There would be no dancing or even a sit-down meal, just picked-over finger foods with a bunch of mining executives, and the other ones. Oh, prospectors, that's what they were called. A tough-looking lot, their faces were always weather-beaten, their hands as rough as sandpaper when she was introduced to some of them and was expected to endure their handshakes.

She hated it. But she was nothing more than a showpiece at these conventions, with their endless table displays of rocks and weird-looking maps hanging everywhere. She didn't understand any of it. Did they really find gold and silver in those godforsaken places? She couldn't imagine being away from her coffee houses and fine eateries, not to mention clean washrooms. Imagine, actually *going* out there in those ugly woods. She shuddered at the unthinkable as she fingered the gold necklace at her throat.

* * *

"You're the buggers who killed off all the Beothuks, aren't you? I just read it in a book about your island."

Kevin Keats was blindsided by the question. He sat across from a potential business partner in a posh restaurant in downtown Toronto. The friend he was having lunch with knew Kevin was of native Mi'kmaq Indian descent. Kevin wasn't laughing. He tried to make light of the comment, though he was burning inside.

"Well, b'y, I wasn't around when all that stuff was goin' on, and I'm not sure if I have to share the blame for what my ancestors did or not. This much I do know, though. If my bloodline were killers of an innocent race of people, that blood has been greatly diluted. My father and uncles and aunts and my grandparents are the gentlest people I know."

His friend apologized, but Kevin dismissed the topic of his ancestry in his easygoing way. The two men spent the next hour discussing minerals, rocks, ore claims, monies that needed to be spent on mineral exploration, and the endless possibilities that Kevin was convinced existed beneath the rough exterior of his home isle.

Kevin had a right to take issue with what his business

associate had joked about. It has been a contentious topic in a few circles for years. A couple of the early European explorers made reference to the possibility that the French had hired the Mi'kmaq nation as allies in their war against the British between the late seventeenth century right up to the beginning of the nineteenth century. French records do not substantiate such a claim. It was only speculation on the part of early observers. Several well-respected authors have researched this theory but could find no evidence to support it. Even the museums of St. Pierre et Miquelon, as well as others in Paris, hold no evidence of the French hiring the Mi'kmaq people to kill the Beothuk. However, there are dozens of recorded accounts of Europeans murdering Beothuk people. If one were to read between the lines of some of these accounts, it is clear that the Mi'kmaq were easy scapegoats.

That same evening found Kevin Keats and his wife, Christina, on the floor of the huge Metro Toronto Convention Centre. The massive hall was a bustle of noise and frenzied activity. The place was filled with so many booths it was a twisted maze of product displays, all of them based on mining, minerals, prospecting, and rare stones and metals. Many of the booths sported multicoloured maps, most of which only a geologist could understand.

The atmosphere was exciting, almost explosive. The cacophony of sound from the floor overpowered first-timers, but soon men and women alike accepted this as normal and blended in with the crowd. For anyone interested in minerals, no matter the field, this was the place to be. For this night, at least, this trendy city facility would host an international family that were all "rock solid."

Somewhere milling through the crowded floor were Kevin's brother, Suley, their father, Allan Keats, and Al's brothers Calvin and Fred. Missing were Al's brother Suley, Sr., and their father, Ted, both of whom were not much into formal affairs. Also part of the same group, and taking all of this in, were representatives of the Stares family: Michael, Steve, and Sandy. They had just been presented with the Bill Dennis Prospectors of the Year Award. The Keats, Stares, and Crocker families had received their coveted plaques from the Prospectors and Developers Association of Canada at the nearby Royal York Hotel.

The family of the old Mi'kmaq explorer Soulis Joe was very well-represented.

One section of the convention showed gold bullion, all under the watchful eyes of several Brinks security guards. Rough rocks embedded with glittering golden flecks revealed nature's treasures on cloth-covered tables. Silver bars that would have been the envy of Soulis Joe sat on a tilted display, surrounded by a bed of colourful crushed sediments. Bronze and coal, along with base metals such as lead, zinc, copper, and iron were available for the viewing. Almost all of the displays came with huge posters explaining the metals' origins.

One of the medium-sized booths caught Kevin's attention. A lifelike picture of an Egyptian queen glared at passersby from a blue velvet background. Probably Cleopatra, Kevin thought as he paused to take in the presentation. A glittering crown, in the very centre of which reared a coiled snake holding a white diamond – Cleopatra's favourite jewel – topped her long forehead. Silver combs held the queen's black hair to either side of her head. Rings of jade pierced her ears.

Her eyes stared out from beneath twin smears of blue-black paint. The woman's neck was covered with an arrangement of garnets and emeralds held in place by a gold string, all weighted down by a small red orb that stopped just above her slight breasts. Her slim, naked arms jutted through several silver and gold amulets. She wore rings that carried rare gems on every finger.

Sitting below this portrait was another lady, her legs crossed to reveal their shape and length. To complement her cleavage, her black evening gown was slit up the crossed leg. Sets of diamond necklaces, bracelets, and rings peered out from plush velvet boxes of green, red, and gold on the wide table beside her. They all looked very expensive. There were no price tags attached to any of the jewels.

Behind the table, and standing as Kevin approached, was her husband. He looked comfortable in his pinstriped three-piece suit. The two men talked about the beauty of the precious stones in front of them. Kevin tried to bring the talk around to a subject more familiar to him, where the wealth had come from. It wasn't easy. This man was into sales and was there to display his wares. He wasn't interested in how they were obtained.

He turned to introduce his wife, almost as an afterthought. Kevin held out his hand. When she allowed her fingers to meet his, they felt clammy, indifferent and very cold. She barely turned her head his way. Kevin kept his thoughts to himself and moved away.

* * *

In general, conventions can go on for hours. It is rare to get much nighttime sleep. Wired after the long day of events, Kevin sat by the window in his hotel room. It was very late. The night's festivities were over. Traffic still streamed along. He thought he could see the first telltale signs of day between the high rises, but he wasn't sure. He wasn't even sure if he was facing east.

For a moment his head was filled with stories from the road that had brought him and his family to this place. He thought of his grandfather, and his father's bold move into mineral exploration, and how his great-great-grandfather, Soulis Joe, had died taking the secret of his lost mine with him. Kevin's own young life was also filled with adventures in the search for rare metals.

A small plane flew into view and disappeared out of his sight, its flashing blue and red lights competing for attention above the buildings. He smiled to himself, his memory borne away with the bush plane. These tough little planes were no strangers to him. He couldn't even remember all of the smooth as well as rough waters all over his own province of Newfoundland and Labrador that he had roared down upon in these aircraft. And even in the northern part of this province, Ontario, where his two cousins Mike and Steve, who were still in the mineral exploration business, now lived. Big airliners, too, had carried him away to places he had only dreamed about. He had searched for minerals in Indonesia, where women of small stature carried steel drill pipes in the sweltering tropical heat. He had been all over North America, as well. All of his relatives had done the same, all in the same field of exploration. He remembered his uncle Fred telling about his trip to Argentina. His method of transportation up into the mountains wasn't a plane on that trip, but a burro.

He remembered another place, in northern Canada. Akimiski Island, the biggest one in James Bay, where he and his first cousins, Steve and Mike, had survived a helicopter crash. Fortunately they were only two hundred feet above ground when it dropped.

They were staking claims for a mining company out in the middle of nowhere. They had finished the site work and had lifted off. The engine "flamed out" and died, fired up again, and then quit. The chopper fell like a spiralling stone. It broke its blades on impact, tore off its skids, and smashed the tail boom. Steve had a slight injury and the others walked away unhurt. He had always heard that any crash you can walk away from is a good one. This was one of them, for sure. They knew there was another chopper working in the same area, so rather than use up their emergency flare, they waited for four hours until it passed them, and they were rescued.

These days Kevin spends most of his time sitting at a desk, dealing options and coordinating projects. Not satisfied with being a field man, he had wanted the challenge of moving on with his own ideas, his own decisions, and his own company. He had wanted to be his own boss.

The boardrooms of mining and exploration investment – most of them on the Canadian mainland – were as tough to navigate as any bogland. Most of the executives were as hard to deal with as the minerals they searched for. There were many days when he wished he were still a hands-on prospector, walking the hidden valleys and the high ridges and struggling across the wet marshes of Newfoundland.

Even being stranded on the tundra of the Canadian Shield had its own rewards.

The sun had come up, but he hadn't noticed it. No breaking over the horizon. No streaming its morning glory over the tall trees. The thousands of city lights below had winked out and the new day was here.

Turning away from the window without closing the curtain, and without looking back, as was his habit, Kevin strode across the thick, carpeted floor, wishing he were in another place, where the sun had long since risen.

Afterword

Benton is not much bigger today than it was in the fifties. It is primarily a bedroom community for the growing town of Gander, just a few minutes' drive west along the Trans-Canada Highway. It is a town that was, in essence, cut out of the Central Newfoundland wilderness.

From the cemetery less than a kilometre from the TCH, where the forest community lay their dead, to the shores of Soulis Pond, where the town was born and the road ends, is no more than two kilometres. Even by Newfoundland standards, an island famous for its many diminutive towns, Benton is indeed small, in size and population. Situated on the south end of the seven-mile-long Soulis Pond, the town lies flat and ends along the one paved road leading to the highway.

When Robert Gillespie Reid was contracted by the Newfoundland government to build a railbed across the island in the year 1890, one of the first problems he encountered was obtaining a good supply of timber. The number of wooden ties needed to lay across and carry the miles of steel rail was enormous. Timbers could be obtained from coastal sawmills and brought to the railheads by schooner, from St. John's in the east, and Port aux Basques in the west. It was very expensive.

Reid wanted unbelievable concessions before he would finish the project. He demanded an agreement which would give him surface and mineral rights to 5,000 acres of land for each mile of rail he operated. This 1893 agreement gave to Reid an amazing 6,500 square miles of some of the best land on the entire island, all of it within easy railway access. Further to this, Reid wanted to operate the finished railway for ten years after its completion, with a generous salary, of course.

Huge tracts of mature forest, beginning at water's edge along the entire length of Soulis Pond and extending far and away beyond its ridges, were a goldmine to the railway engineers. Here was an unlimited supply of timber that was so vast it could build a railbed ten times longer than the one they required.

A sawmill was started on the shores of the pond in 1894, and until the railway's completion in 1898, Benton would be a major lumber depot for the "Reid" railway. This timber supply kept the town alive on into the twentieth century, when the pulp and paper company, Bowaters of Corner Brook, maintained the town as its main depot for dozens of their logging camps. All of these depended on the steady influx of men and materials that detrained at Benton and were portaged down Soulis Pond by boat in spring and over its frozen breast in winter. The town has volumes of forest tales yet to be told.

Not only did the forest resources around the Benton area

of Soulis Pond provide easy access to timbers needed for the railbed construction, it also provided another much-needed material. A strong foundation was required before the timbers could be laid. Along the entire proposed route, marshlands and tundra-like areas challenged the rail path.

Granite, the best of all foundations, was the answer. Benton was the site of Newfoundland's first known quarry. Great volumes of the rock were blasted out of the nearby outcrops and would fortify the railway construction for miles beyond the new town.

It is ironic that the town had part of its beginning in the mining business, for its very name is derived from a mineral. A sticky, versatile clay can be found in several places around the area. Bentonite is a rare, absorbent form of clay. It is finely grained and will swell to several times its normal volume when saturated with water. It is used in the petroleum industry as an oil filter, and also as a sealant in drill holes to prevent collapse while drilling. Bentonite is a residue of decomposed rock formed from a volcanic ash. It seems that though the town is young, it is sitting on very old roots.

Benton continues to play a role in the forest industry. It is still well-known for its hunting and fishing. However, it is now much better known for its family of prospectors. Although this book has focused on the Keats family of prospectors from Benton and has carried the influence of Allan Keats throughout its pages, it must be made clear that the Keats name is not the only one connected with the prospecting and mining industry in that town. There are other names. The Stareses, Crockers, Smiths, and Barretts are prospectors all. They all have one common thread. The blood of the old Mi'kmaq Soulis Joe forever courses through their veins. They all have their stories to tell, though many of them seldom talk about their feats. It is also a difficult story to tell, for it is still ongoing, a never-ending volume of discoverers of the earth's valuables.

Benton Resources, a mineral exploration company founded by Stephen and Michael Stares, was the first of Soulis Joe's descendants' to be publicly traded. While this book was in progress, Kevin Keats, Allan's son and the president of a prospecting exploration company called Golden Dory, announced to the world its new and major discovery of an ore body rich in uranium on the south coast of this province. As of September 8, 2008, this company is also listed on the TSX Venture Exchange – to the tune of $1.8 million – as specializing in the search for uranium, gold, and base metals anywhere in Canada, but especially in the company's home province of Newfoundland and Labrador.

The name Golden Dory must have been well thought out. "Dory" is taken from the ancient word *dore*, meaning gold bullion. It also refers to a mixture of gold and silver bars and is pronounced the same as the Newfoundland boat of the same name.

Shortly after this announcement, Sandy Stares became newsworthy as having discovered one of the first platinum deposits on the island of Newfoundland. Just before this manuscript was sent off to my publisher, Kevin Keats was again headlined in the papers. He predicted another major gold find would be announced from the Central Newfoundland area. This family will continue to surprise the world with its findings.

This story has centred around the Keats family because of my involvement with Allan Keats during a much-loved summer's employment in the field of prospecting. I will never forget it.

Acknowledgements

Without Kevin Keats's request for me to write this book, it would not have been done. At least not by me. I knew nothing about Soulis Joe, though I always wondered where the name on the highway sign just east of Gander came from. Thanks to Kevin Keats, I now know. In the search for Soulis Joe's story, which took me all over this island, I found not only the deep-rooted history of an ancient people, but also a wonderful legacy. The soft voice of Allan Keats, answering my countless questions while I was seated in the tree shade along some nameless brook, will stay with me forever. Thanks, Al. There may be some places in these pages where the reader is taken that he or she may doubt. But in the freedom of my mind, I believe. Away from the beaten path, where the true story began, I found and will continue to follow a yearning of my own. Thank you both.

Thanks to my first editor, constant sounding board, best love and friend, Rose. And thank you to Clint Collins, my favourite artist and photographer – my son.

Also, a special thanks to Garry, Jerry, and especially Margo Cranford of Flanker Press.

(Illustration by Clint Collins)

Allan Keats, west of Birchy Lake, summer 2008. (Author photo)

Daniel Weidendorfer, Allan Keats, and Julie Brown on the Baie Verte Peninsula, summer 2008. (Author photo)

Allan Keats searching from the canoe at Red Indian Lake, summer 2008. (Author photo)

Allan Keats at the west end of Red Indian Lake, summer 2008. (Author photo)

Allan Keats inspecting a rock with his lens, Southwest Gander River, summer 2008. (Author photo)

Allan Keats standing next to a rock too big to bang, La Scie road, summer 2008. (Author photo)

Roughing it near Birchy Lake, summer 2008. (Author photo)

Allan Keats resting, somewhere south of Red Indian Lake, summer 2008. (Author photo)

Allan Keats, great-grandson of Soulis Joe. Panning gold in 2002, location undisclosed. (Photo courtesy of Kevin and Christina Keats)

Allan Keats in a cave somewhere on the Hampden highway, summer 2008. (Author photo)

Kevin Keats prospecting on the Burin Peninsula in 2008. He is at the mouth of an old silver adit from the 1800s. Kevin is a great-great-grandson of Soulis Joe. (Photo courtesy of Kevin and Christina Keats)

Allan and Kevin Keats. Photo taken in May of 2007 during the opening of the Duck Pond copper, lead, and zinc mine in Central Newfoundland. Allan was instrumental in this discovery. (Photo courtesy of Kevin and Christina Keats)

Kevin Keats prospecting in Northern Ontario, 2001. (Photo courtesy of Kevin and Christina Keats)

Kevin Keats prospecting in Northern Ontario, 2001. (Photo courtesy of Kevin and Christina Keats)

Kevin Keats prospecting in Mexico, 2007. (Photo courtesy of Kevin and Christina Keats)

Sandy Stares and Kevin Keats in Mexico, 2007. (Photo courtesy of Kevin and Christina Keats)

Calvin Keats holding a snake while prospecting in the jungles of Sumatra in Indonesia in 1997. Calvin is a great-grandson of Soulis Joe. (Photo courtesy of Kevin and Christina Keats)

Calvin Keats (right) and his stepson, Deon Humby. (Photo courtesy of Kevin and Christina Keats)

Calvin Keats, second from left, prospecting in Sumatra, Indonesia, in 1997, with local Indonesian helpers. Mount Kerenchi, an active volcano, is shown in the background. (Photo courtesy of Kevin and Christina Keats)

Fred Keats prospecting in the Andes Mountains of Argentina in 2002. (Photo courtesy of Kevin and Christina Keats)

Fred Keats prospecting in Africa in 2007. Fred is a great-grandson of Soulis Joe. (Photo courtesy of Kevin and Christina Keats)

Fred Keats prospecting in the Arctic in 1995. (Photo courtesy of Kevin and Christina Keats)

Wesley Keats prospecting in Biset, Manitoba, in 2007. Wesley is Soulis Joe's great-great-grandson. (Photo courtesy of Kevin and Christina Keats)

Theodore "Ted" Keats, Soulis Joe's grandson. This photo was taken in the Millertown area in 1975 during his prospecting days with Noranda Exploration. (Photo courtesy of Kevin and Christina Keats)

Alexander Stares, Stephen Stares, Michael Stares, and their father, Robert Stares. Taken while prospecting in the Dawson Township gold occurrence in Northern Ontario in 1998. Robert is married to Soulis Joe's great-granddaughter, Effie Stares (née Keats). (Photo courtesy of Kevin and Christina Keats)

The Stares boys. L-R: Robert, Alexander, Stephen, Michael. (Photo courtesy of Kevin and Christina Keats)

Alexander Stares (Soulis Joe's great-great-grandson) with his son, Shane Stares, the youngest member of this prospecting family, taken at H-Pond near Gander in 2003. Shane found the first gold-bearing rock at H-Pond. (Photo courtesy of Kevin and Christina Keats)

L-R Front Row: Fred Keats, Calvin Keats, Allan Keats, Robert Stares (Soulis Joe's great-grandsons). L-R Back Row: Shane Stares (Soulis Joe's great-great-great-grandson), Deon Humby, Alexander Stares, Michael Stares, Stephen Stares, Kevin Keats, Wesley Keats. (Photo courtesy of Kevin and Christina Keats)

L-R Front Row: Deon Humby, Shane Stares, B. J. Bursey, Suley Keats, Jr., Wesley Keats. **L-R Back Row:** Alexander Stares, Allan Keats, Rick Crocker, Robert Stares, Michael Stares, Kevin Keats, Fred Keats, Stephen Stares, Calvin Keats.

This photo was taken at the Prospectors and Developers Association of Canada (PDAC) Convention in 2008. The Keats and Stares families were the recipients of the 2007 Bill Dennis Prospector of the Year Award. It was awarded for a lifetime of achievement and contribution to the mineral industry: five generations of active prospectors. (Photo courtesy of Kevin and Christina Keats)

Cecil Barrett, great-great-grandson of Soulis Joe. (Photo courtesy of Kevin and Christina Keats)

Winston Barrett (deceased), great-great-grandson of Soulis Joe. (Photo courtesy of Kevin and Christina Keats)

Brian Bursey, Jr., great-great-great-grandson of Soulis Joe, and Allan Keats's grandson. (Photo courtesy of Kevin and Christina Keats)

L-R: Ted, Edgar, Jerry, and Ray Smith, all great-great-grandsons of Soulis Joe. (Photo courtesy of Kevin and Christina Keats)

Edgar, Jerry, Ray, and Ted Smith with their mother, Geraldine Smith, great-granddaughter of Soulis Joe. (Photo courtesy of Kevin and Christina Keats)

Suley Keats, Jr., great-great-grandson of Soulis Joe, and Suley, Sr., great-grandson of Soulis Joe. (Photo courtesy of Kevin and Christina Keats)

Calvin Crocker, back row centre. Calvin is married to Madeline Keats, great-granddaughter of Soulis Joe. (Photo courtesy of Kevin and Christina Keats)

Rick Crocker, son of Calvin Crocker. (Photo courtesy of Kevin and Christina Keats)

Kevin Keats and son Stephen. (Photo courtesy of Kevin and Christina Keats)

Left: Soulis Joe. (Photo from *Nestled Between the Hills* by Arthur E. Penney) **Right:** Matthew "Matty" Mitchell. (Eric Swanson photo. Courtesy of the Newfoundland and Labrador Heritage website)

Bibliography

Anger, Dorothy. *Noywa'mkisk (Where the Sand Blows . . .): Vignettes of Bay St. George Micmacs.* Port au Port: Bay St. George Regional Indian Band Council, 1988.

Barnes, Michael. *More Than Free Gold: Mineral Exploration in Canada Since World War II.* Renfrew, ON: General Store Publishing House, 2008.

Bercuson, David J., and J. L. Granatstein. *The Collins Dictionary of Canadian History: 1867 to the Present.* Toronto: HarperCollins Canada, 1988.

Blackwood, R. F. "Geology of the east half of the Gambo (2D/16) map area and the northwest portion of the St. Brendan's (2C/13) map area, Newfoundland." Report 77-5. St. John's: Mineral Development Division, Department of Mines and Energy, Government of Newfoundland and Labrador, 1977.

Coish, Calvin, ed. *Stories of the Mi'kmaq*. Grand Falls: College of the North Atlantic, 2000.

Cooper, Afua. *The Hanging of Angelique: The Untold Story of Canadian Slavery and the Burning Of Old Montreal.* Toronto: HarperCollins Canada, 2006.

Dow, George Francis. *Slave Ships and Slaving*. Cambridge, MA: Cornell Maritime Press, 1968.

English, George Allen. *The Greatest Hunt in the World.* Montreal: Tundra Books, 1969.

Gibbons, R. V., and N. Mercer. *Mining in Newfoundland and Labrador: From 1949 to 1982.* Montreal: The Canadian Institute of Mining and Metallurgy Directory, 1982.

History of Halifax, Nova Scotia. http://www.halifaxkiosk.com/history.php

Howley, James P. *The Beothuks or Red Indians: The Aboriginal Inhabitants of Newfoundland.* New York: AMS Press, 1979.

Jackson, Doug. *On the Country: The Micmac of Newfoundland.* St. John's: Harry Cuff Publications, 1993.

Martin, Wendy. *Once Upon a Mine: Story of Pre-Confederation Mines on the Island of Newfoundland.*

Montreal: The Canadian Institute of Mining and Metallurgy, 1983.

Murray, Alexander, and James P. Howley. *Geological Survey of Newfoundland.* London: Edward Standford, 1881.

Pastore, Ralph T. *The Newfoundland Micmacs: A History of Their Traditional Life.* St. John's: Newfoundland Historical Society, 1978.

Paul, Daniel N. *First Nations History: We Were Not the Savages.* 3rd ed. Halifax: Fernwood Publishing, 2006.

Penney, Arthur E. Nestled *Between the Hills: A History of Cannings Cove.* St. John's: Jesperson Press, 1991.

Peters, H. R. "The History of Mining in Newfoundland, 1857–1949." *Newfoundland Journal of Geographical Education* 5, no. 2 (1980): 1.

Pulling, George C., and Ingeborg C. Marshall. *Reports and Letters: Relating to the Beothuk Indians of Newfoundland.* St. John's: Breakwater Books, 1989.

Ryan, B., R. J. Wardle, C. F. Gower, and G. A. G. Nunn. "Nickel-Copper Sulphide Mineralisation in Labrador: The Voisey's Bay Discovery and Its Exploration Implications." Current Research, Report 95-1. St. John's: Geological Survey, Department of Natural Resources, Government of Newfoundland and Labrador, 1995.

Service, Robert. *Dan McGrew, Sam McGee: The Poems of Robert Service.* New York: Barnes and Noble Books, 1987.

Taylor, S. W., S. J. O'Brien, and H. S. Swinden. *Geology and Mineral Potential of the Avalon Zone and Granitoid Rocks of Eastern Newfoundland.* St. John's: Mineral Development Division, Department of Mines and Energy, 1979.

Tompkins, Edward. *Ktaqmkukewaq Mi'kmaq: Wlqatmuti / The Mi'kmaw People of Newfoundland: A Celebration.* Corner Brook: Federation of Newfoundland Indians, 2004.

Traditional Mi'kmaq Culture: Aboriginal peoples: Newfoundland and Labrador Heritage. http://www.heritage.nf.ca/aboriginal/mikmaq_culture.html

Whitby, Barbara. *The Last of the Beothuk: A Canadian Tragedy.* Toronto: James Lorimer & Co., 2005.

White, James, ed. *Handbook of Indians of Canada.* Published as an Appendix to the Tenth Report of the Geographic Board of Canada, Ottawa, 1913.

Worcester, E. *Life's Adventure: The Story of a Varied Career.* New York: Charles Scribner's Sons, 1932.

Young, Don. "A History of St. Paul Island: From the Memories of Lena Campbell (Jones)." St. Paul Island.

http://www.geocities.com/Heartland/Estates/6001/St_Paul_Island.htm.

Other Credits:

Boyd's Cove Beothuk Interpretation Centre

Bull and Bear

CIM Magazine

Donna Pardy, Town of Gambo

Dover Fault Interpretation Site

Federation of Newfoundland Indians

Gros Morne National Park

Louis Collins, Hare Bay

Marianopolis College

Marine Studies, Marine Institute, Memorial University

Matty Mitchell Prospectors Resource Room

Mi'kmaq Resource Centre, Cape Breton University

Baie Verte Peninsula's Miner's Museum

The Northern Miner newspaper

The Rooms

Theophilus Collins, Hare Bay

Tom Burry, Department of Natural Resources, Gambo

Twin Mining Corp Annual Report

About the Author

Gary Collins was born in a small, two-storey house by the side of the sea in the town of Hare Bay, Bonavista North. He finished school at Brown Memorial High in the same town and resides in Hare Bay still.

He spent forty years in the logging and sawmilling business along with his father, Theophilus, and son Clint. Gary was once Newfoundland's youngest fisheries guardian. He managed log drives for years down spring rivers, spent seven seasons driving tractor trailers over ice roads and the Beaufort Sea of Canada's Western Arctic, and has been involved in the crab, lobster, and cod commercial fisheries. His writing career began when he was asked to write eulogies for deceased friends and family. He spent a full summer employed as a prospector before he wrote *Soulis Joe's Lost Mine*. He liked the work so much, he went back to school to earn his prospecting certificate.

His favourite pastimes are reading and playing guitar at his log cabin, and now, "banging rocks." Gary Collins is married to the former Rose Gill, and together they have three children and three grandchildren.

Soulis Joe's Lost Mine is Gary Collins's third book. His first two books, *Cabot Island* and *The Last Farewell*, are Newfoundland bestsellers.

ALSO BY GARY COLLINS

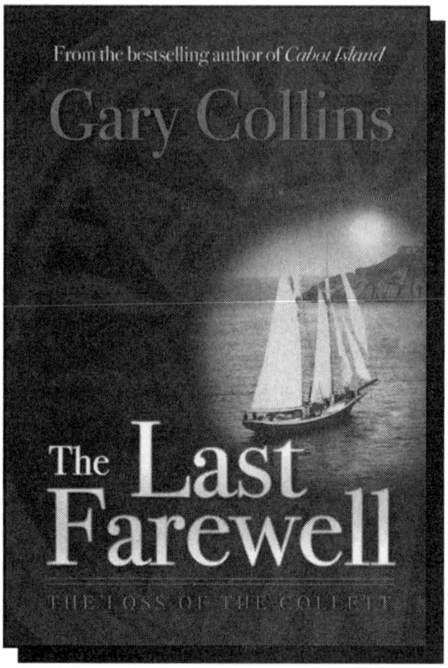

"Read *The Last Farewell* not only because it is a moving historical tale of needless tragedy but also because it's a book enriched with abundant details of Newfoundland life not so widespread anymore."
The Pilot

The Last Farewell tells the true story of a crew of logger-sailors who left their home port of Hare Bay aboard a two-masted schooner in early June of 1934. Along her route to St. John's, the crew of the *Ethel Collett* tell each other stories of life and death on the sea. They relive some of Newfoundland's richest historical moments, from shipwrecks and sealing disasters to political strife and financial ruin. But little do they know that they are heading toward one of the most astonishing tales of them all: their own.